Integration of GIS and Remote Sensing

Mastering GIS: Technology, Applications and Management Series

Integration of GIS and Remote Sensing

Edited by

Victor Mesev

Department of Geography, Florida State University, USA

BICENTENNIAL
1807
WILEY
2007
BICENTENNIAL

John Wiley & Sons, Ltd

Other Wiley Editorial Offices

John Wiley & Sons Inc., 111 River Street, Hoboken, NJ 07030, USA

Jossey-Bass, 989 Market Street, San Francisco, CA 94103-1741, USA

Wiley-VCH Verlag GmbH, Boschstr. 12, D-69469 Weinheim, Germany

John Wiley & Sons Australia Ltd, 42 McDougall Street, Milton, Queensland 4064, Australia

John Wiley & Sons (Asia) Pte Ltd, 2 Clementi Loop #02-01, Jin Xing Distripark, Singapore 129809

John Wiley & Sons Canada Ltd, 6045 Freemont Blvd, Mississauga, Ontario L5R 4J3, Canada

Wiley also publishes its books in a variety of electronic formats. Some content that appears in print may not be available in electronic books.

Anniversary Logo Design: Richard J. Pacifico

Library of Congress Cataloging in Publication Data

Integration of GIS and remote sensing / edited by Victor Mesev.
   p. cm. — (Mastering GIS : technology, applications and management series)
  Includes bibliographical references and index.
  ISBN 978-0-470-86409-8 (cloth : alk. paper)
  1. Geographic information systems. 2. Remote sensing. I. Mesev, Victor.
  G70.212I573 2007
  910.285—dc22   2007029099

British Library Cataloguing in Publication Data

A catalogue record for this book is available from the British Library

ISBN 978-0-470-86409-8 HB
ISBN 978-0-470-86410-4 PB

Typeset in 10/12pt Times by Integra Software Services Pvt. Ltd, Pondicherry, India
Printed and bound in Great Britain by TJ International, Padstow, Cornwall
This book is printed on acid-free paper.

Contents

3   Data fusion related to GIS and remote sensing

Paolo Gamba and Fabio Dell'Acqua

4   The importance of scale in remote sensing and GIS and its implications for data integration

Peter M. Atkinson

5   Of patterns and processes: spatial metrics and geostatistics in urban analysis

XiaoHang Liu and Martin Herold

8   Integrating remote sensing, GIS and spatial modelling for sustainable urban growth management   173
Xiaojun Yang

9   An integrative GIS and remote sensing model for place-based urban vulnerability analysis   199
Tarek Rashed, John Weeks, Helen Couclelis and Martin Herold

10   Using GIS and remote sensing for ecological mapping and monitoring   233
Jennifer A. Miller and John Rogan

Series Foreword

Since 2001 it has been my privilege to be involved with the book *Geographic Information Systems and Science*, published by John Wiley and Sons, Ltd. Through its various editions, this book and associated materials has sought to present a state-of-the-art overview of the principles, techniques, analysis methods and management issues that come into play whenever the fundamental question 'where?' underpins decision-making.

Together this material makes up the organising concepts of Geographic Information Systems (GIS), which has a rich and varied history in environmental, social, historical and physical sciences. We can think of GIS as the lingua franca that builds upon the common purposes of different academic traditions, but with an additional unique emphasis upon practical problem-solving. As such, much of the core of GIS can be thought of as transcending traditional academic disciplinary boundaries, as well as developing common approaches to problem solving amongst practicing professionals.

Yet many of the distinctive characteristics, requirements and practices of different applications domains also warrant specialised and detailed treatments. 'Mastering GIS' seeks to develop and extend our core understanding of these more specialised issues, in the quest to develop ever more successful applications. Its approach is to develop detailed treatments of the requirements, data sources, analysis methods and management issues that characterise many of the most significant GIS domains.

First and foremost, this series is dedicated to the needs of advanced students of GIS and professionals seeking practical knowledge of niche applications. As such, it is dedicated making GIS more efficient, effective and safe to use, and to render GIS applications ever more sensitive to the geographic, institutional and societal contexts in which it is applied.

Paul Longley, Series Editor
Professor of Geographic Information Science
University of London

Preface

'Integration' refers to the move towards a closer, perhaps symbiotic, relationship between geographic information systems (GIS) and remote sensing and is seen by many to be essential for the future development of both technologies. The main drivers for this move have been the proliferation of geospatial data in various formats, the pursuit of sophisticated statistical models, and the demand for more elaborate applications. Moreover, proprietary systems are no longer solely devoted to either GIS operations or image processing; all now handle data from both and all now offer analytical functions that facilitate dual interoperable analyses.

Ever since the first formal research agendas on GIS and remote sensing integration were introduced, back in 1990, by the US National Center for Geographic Information and Analysis (NCGIA Initiative I-12), hybrid systems have become very much the rule rather than the exception. Although data from both GIS and remote sensing are now routinely analysed by seamless interoperable amalgamated systems, many users are unaware of the numerous technical and institutional issues that need to be addressed when merging data that are derived from disparate sources and essentially represent diametrically opposing conceptual views of reality. This book explores the tremendous potential that lies along the interface between GIS and remote sensing for activating seamless databases and instigating information interchange. It concentrates on the rigorous and meticulous aspects of analytical data matching and thematic compatibility – the true roots of all branches of GIS–remote sensing applications. The first four chapters of the book confront technical issues of integration, such as data fusion, scale effects and data uncertainty, as well as introducing an integrated taxonomy of data structure and system-independent functionality. The remaining chapters explore and demonstrate most of the salient integration procedures and methodologies, using a number of applications, including the measurement of urban morphologies, the estimation of urban sprawl and population growth, urban vulnerability analysis, and the augmentation of environmental change indicators. In all, emphasis is given to the close statistical and thematic association of information from both technologies, and the merits of joint implementation of GIS and remote sensing.

This book is the result of an extensive research by experts working at the interface of GIS and remote sensing and will appeal to students and professionals dealing with not only GIS or remote sensing but also computer science, civil

engineering, environmental science and urban planning in the academic, govern-
mental and commercial/business sectors. The Editor is grateful to all the authors and
anonymous reviewers for their time and effort and for keeping to a strict schedule.
Acknowledgements are also due to staff at Wiley, especially Fiona Woods and
Lyn Roberts, for their patience and encouragement, to Jennifer Miller, Kimberly
Hattaway, Aaron Binns, Michael Sims (design cover) and Charles Layman for
constructive advice, and to Alexandra Walrath, who painstakingly (and enthusiasti-
cally) proofread the entire manuscript, contributed to the Introduction and compiled
the Index. Lastly, the editor would like to dedicate this book to Lady Hilda and Sir
Geoffrey Miller, who provided guidance and jovial support in abundance.

Victor Mesev
Tallahassee, FL, USA
February 2007

List of Contributors

Peter M. Atkinson School of Geography, University of Southampton, UK

Helen Couclelis Department of Geography, University of California at Santa Barbara, CA, USA

Fabio Dell'Acqua Telecommunications and Remote Sensing Laboratory, Universita Degli Studi Di Pavia, Italy

Manfred Ehlers Forschungszentrum für Geoinformatik (FZG) und Fernerkundung, Universität Hochschule Vechta, Germany

Paolo Gamba Telecommunications and Remote Sensing Laboratory, Universita Degli Studi Di Pavia, Italy

John Hasse Department of Geography and Anthropology, Rowan University, Glassboro, NJ, USA

Martin Herold Institut für Geographie, Friedrich-Schiller-Universität Jena, Germany

XiaoHang Liu Department of Geography and Human Environmental Studies, San Francisco State University, CA, USA

Victor Mesev Department of Geography, Florida State University, Tallahassee, FL, USA

Jennifer A. Miller Department of Geography and the Environment, University of Texas at Austin, 1 University Station – A3100, Austin, TX 78712-5116, USA

Tarek Rashed Department of Geography, University of Oklahoma, Norman, OK, USA

John Rogan Graduate School of Geography, Clark University, 950 Main Street, Worcester, MA 01610, USA

Tara Shine Environment and Development Consultant, 127 The Meadows, Belgooly, Co. Cork, Ireland

Alexandra Walrath   Department of Geography, Florida State University, Talla-
                        hassee, FL, USA

John Weeks   Department of Geography, San Diego University, CA,
                        USA

Changshan Wu   Department of Geography, University of Wisconsin at
                        Milwaukee, WI, USA

Xiaojun Yang   Department of Geography, Florida State University, Talla-
                        hassee, FL 32306, USA

1

GIS and remote sensing integration: in search of a definition

Victor Mesev and Alexandra Walrath

Department of Geography, Florida State University, Tallahassee, FL, USA

> *Synergy – the bonus that is achieved when things work together harmoniously* – Mark Twain

> *Wisdom implies a mature integration of appropriate knowledge, a seasoned ability to filter the inessential from the essential* – Deborah Rozman

1.1 Introduction

Ever since its formalism by the NCGIA Initiative-12 in 1990 (Star *et al.*, 1991), the move towards 'seamless' and 'hybrid' integration of data, techniques and organization from the geographic information systems (GIS) domain with those from the remote sensing[2] sphere has been arduous, sporadic and irresolute. Few major breakthroughs have materialized other than the establishment of routine data format interchanges, improvements in the efficiency of interoperational relational database systems (Abel *et al.*, 1994) and modest advances in the accuracy of object/thematic

---

[1] GIS is used both singularly and as a collective throughout this book. GIS is typically defined as 'a computer system for the collection, storage, manipulation, display and management of spatial information'.

[2] Remote sensing predominantly refers to the collection and manipulation of digital satellite imagery.

Integration of GIS and Remote Sensing   Edited by Victor Mesev

identification cross-overs (Shi *et al.*, 1999). More ambitious endeavours to create truly *integrated* geographic information systems (IGIS), sometimes called 'total' integration, seem to have floundered on most of the initial conceptual, technical and institutional obstacles identified by the NCGIA initiative (cf. Ehlers, 1989; Star *et al.*, 1990; Hinton, 1996; Wilkinson, 1996; Mesev, 1997). One could even say that the search for more resolute solutions, such as those related to the object/field dichotomy, analytical interoperability, the close monitoring of error propagation and the compatibility of mutually beneficial research programmes, remains as elusive today as it was in 1990. Admittedly, many proprietary geospatial systems are capable of representing and querying data stored in an increasing number of formats and resolutions, yet computational compatibility is rarely translated to full conceptual, thematic, scale and temporal compatibility. In other words, although technical expediency has facilitated the handling of data from GIS and remote sensing, there is no guarantee that any subsequent computational interaction necessarily results in strong intuitive and theoretical mutual relationships. Total integration may not be a question of whether GIS and remote sensing *can* be integrated, but more of whether they *should* be integrated – and to answer that, some discussion is first required on precisely what integration between GIS and remote sensing actually means.

1.2   In search of a definition

No one definition of integration between GIS and remote sensing exists. Instead, integration has been used to refer indiscriminately to almost any type of connection, ranging from pragmatic computational amalgamation of data to the conceptual understanding of how geographic features are interrelated. Unsurprisingly, an unbounded definition embraces a large and growing body of literature, anything from research on tight, seamless databases, and robust statistical relationships (Zhou, 1989; Smits, 1999), to applications of variable implicitness and unpredictable levels of information exchange (cf. de Brouwer *et al.*, 1990; Janssen *et al.*, 1990; Davis *et al.*, 1991; Chagarlamundi and Ganulf, 1993; Debinski *et al.*, 1999; Driese *et al.*, 2001; Brivio *et al.*, 2002). However, in the search for a narrower definition, any book with the term 'integration', to all intents and purposes, presumes a strict discussion on numerical calculations and complex computational algorithms, especially when the integration is referring to system-based technologies, such as GIS and remote sensing. In this sense, integration may be defined as the establishment of numerical consistency across disparate digital data models and the execution of robust programming algorithms (Archibald, 1987; Brown and Fletcher, 1994; Abel *et al.*, 1994). In addition, emphasis is on computational schemata that ensure either efficient dual operability across software platforms or, preferably, the creation of a hybrid database capable of handling incongruent data at variable resolution, complexity, quality and completeness (Zhou, 1989). Under this definition, the integration of data (the beginnings of data fusion) and algorithms may be numerically and operationally feasible, but it does not necessarily cover the blending of disparate data and algorithms pertinent to information that is explicitly *geographic* in nature. The jump from generic numerical data to geographic data represents more than simply adding a locational dimension. Both the quality and usefulness of spatial data that represent and model the complex real world are intrinsically constrained by three basic cartographic rules: the scale of representation; the generalization of feature delineation; and the semantic description of parcels of the Earth's surface and atmosphere. These three conditioning factors are further intensified by the eternal pursuit for greater accuracy and higher precision when recording the exact locational coordinates of geographic features.

Both GIS and remote sensing are technologies that focus exclusively on geographic data and, as such, both are designed to represent the world's geographic features as reliably and realistically as possible and within the constraints of the three cartographic rules. However, that is where the straightforward comparison ends. Technically and conceptually, each technology[3] is founded on diverging principles, where remote sensing is predominantly a data collection technology, while GIS is one that is principally dedicated to data handling. Remote sensing deals with the more immediate access of primary data at a more continuous scale, collected over extensive areas at rapid temporal frequencies. Digital remotely sensed data records the magnitude of passive and active energy at multiple wavelengths as it interacts with the earth's surface and atmosphere. As such, remotely sensed pixels are a multispectral radiometric vector that represents the continuous nature of the biophysical and anthropogenic landscapes at various levels of spatial, spectral and temporal resolution. The resultant raster image of individual pixels shows how the landscapes would appear from an elevated viewpoint. However, the image does not have an interrelated topology and the pixels are not implicitly related, other than by their positional adjacency. The continuous representation of reality and the lack of a coherent topology invariably limit the extent to which pure thematic information can be extracted, and as such the accuracy of an image is highly unpredictable, both spatially and thematically.

In comparison, data handled by GIS are commonly stored as vector models and represent geographic features as more discrete entities within a structured topology and defined by implicit relationships. As a result, discrete entities are delineated by sharper, crisp boundaries and labelled with less ambiguous thematic descriptions. However, much of the digital spatial data stored in a GIS are derived from external sources, such as analogue maps, ground surveys, global positioning systems (GPS) and, most importantly, remote sensing (Gao, 2002; Xue *et al.*, 2002). Furthermore, remote sensing, in the form of aerial photographs, is also the predominant resource

---

[3] GIS and remote sensing are referred to as technologies in this book although the terms 'field' or 'discipline' (as incorporated by GIScience) are sometimes used by other sources to indicate broader theoretical underpinnings.

for producing many of the topographic compilations from which environmental indicators, such as elevation, hydrology and land cover, are digitized into sharp vector boundaries and entered into GIS (Dobson, 1993). More recently, satellite remote sensors with high spatial resolutions of 4 m and finer are also providing valuable input data into many GIS applications, especially for the much neglected field of monitoring urban morphologies, urban pollution and urban growth (Mesev, 2003). The traditional role and reliance on remote sensing as input data for GIS suggests that integration is not new and has existed as long as both technologies (Marble, 1981; Piowowar *et al.*, 1990; Wilkinson, 1996). The three time-honoured ways in which GIS and remote sensing have been integrated are as follows:

- *Remote sensing used to collect data for GIS databases.* This includes the ability to update and validate thematic coverages, using aerial photographs, earth observation satellite sensors, interferometric radar and LiDAR.

- *GIS data used as ancillary information for image processing.* Many techniques exist, such as using vector lines to define boundaries between land covers, providing locational attributes for geo-registration, and aiding classification by selecting purer training samples, weighting discrimination functions and sorting classified pixels (see Hutchinson, 1982; Foody, 1988; Mesev, 1998, 2001).

- *Combined analytical functions.* These include basic spatial queries, the overlay of statistical and thematic attributes from both GIS and remote sensing, using Boolean and fuzzy logic, and the building of multiple-view expert systems.

All three of these traditional means of integration were established well before the NCGIA initiative of 1990. According to the initiative, the next step for greater assimilation between GIS and remote sensing depended on greater computer processing power (Faust *et al.*, 1991), reduction in error propagation (Lunetta *et al.*, 1991), compatibility of data structures (Ehlers *et al.*, 1991), and resolution of many non-analytical institutional impediments, such as data availability, costs, standards and organizational infrastructure (Lauer *et al.*, 1991). Unfortunately, the volume of subsequent research has not matched the same sense of importance and urgency expressed by these and other calls to ensure tighter integration.

1.2.1 Evolutionary integration

For some, complete or total integration between GIS and remote sensing is the ultimate goal. Ehlers *et al.*(1989) proposed three stages in the evolution of integration that focused on the degree of interaction between data models, the level of data exchange, the pursuit of close geometric registration, the matching of cartographic representation, a parallel user interface, and the compatibility of geographic abstraction. The three stages of the evolution are as follows:

- *Stage 1* would focus on the separate but equal development of databases from each technology. Data would be exchanged in predominantly vector format (for GIS) and raster models (for remote sensing) but capable of being simultaneously displayed by overlays. Analysis would be limited to the update of GIS coverages by the positional comparison of thematic attributes generated from classified remotely sensed images; or the use of GIS data for facilitating image geo-registration.

- *Stage 2* oversees the continuation of separate databases, but each technology would share a user interface. Data from each technology would be converted to the other through vectorization and rasterization, and the operational rationalization of spatial and temporal attributes.

- *Stage 3* represents the final level of complete or 'total' integration. Essentially, GIS and remote sensing become one indistinguishable system, in which raster and vector data models are handled interchangeably through data uniformity across object-based (GIS data) and field-based (remotely sensed data) geographic representation.

Total integration, although theoretically desirable, is not replicated pragmatically. Instead, much research and applications involving the integration of GIS and remote sensing seems to be adequately completed by stages 1 and 2.

1.2.2 Methodological integration

The three stages in the evolution of integration of data and computational analysis between GIS and remote sensing also presuppose a methodological continuum; generally from loose data coupling to indistinguishable models of representation. However, the continuum is unstructured and integration issues are sporadic and unfocused. Mesev (1997) outlined a logical and structured, yet flexible, frame-work or schema for the formalization of methodological factors and issues for consideration when tackling integration between GIS and remote sensing. The reasons for designing a formal schema were primarily to define all conceivable steps within a structure that defines data accumulation, processing, and decisions in a general chronological order, and also to promote awareness and stimulate discussion of the many pitfalls surrounding the delicate interface between GIS and remote sensing. Organized into a series of hierarchical levels, the top-down approach of the schema ensures that all methodological issues are addressed at increasing detail. Level 1 contains the broadest set and includes data unity, measure-ment conformity, positional integrity, statistical relationships, and classification compatibility – as well as integration design with reference to many non-analytical and external factors such as feasibility and cost–benefit studies. At level 2, links between the six level 1 components become more complex, and by level 3 they

Level II	Level III
Data unity (factors that bring together GIS and remotely sensed data)	
Information interchange	Definition of integration, type of information, information harmony (spatial units and attributes)
Data availability	Awareness, publicity, search, data type, age, quality, (access or create)
Data accessibility	Cost, agreements, exchanges, sharing, proprietary, resistance, confidentiality, liability
Data creation	Digitising, scanning, survey information encoding, sampling, data transformation, GPS
Measurement conformity (factors that link GIS and remotely sensed data)	
Data representation	Data structures (vector, raster), data type, level of measurement, field-based vs. object-based modelling, interpolation
Database design	Type (relational, hybrid), schema, data dictionary, implementation (query, testing)
Data transfer	Format, standards, precision, accuracy
Positional integrity (factors that spatially coordinate GIS and remotely sensed data)	
Generalization and scale	Spatial resolution, scale, data reduction and aggregation, scale invariance
Geometric transformation	Rectification, registration, resampling, coordinate system, projection, error evaluation
Statistical relationships (factors that measure links between GIS and remote sensing)	
Vertical	Boolean overlays, fuzzy overlays, dasymetric mapping, areal interpolation, linear and non-linear equations, time series, change detection
Lateral	Spatial searches, proximity analysis, textural properties
Classification compatibility (factors that harmonize information between GIS and remote sensing)	
Semantics	Classification schemata, levels, descriptions, class merging, standardization
Classification	Stage (pre-, during, post-), level (pixel, sub-pixel) type (per-pixel, textural, contextual, neural nets, fuzzy sets), change detection, accuracy assessment

Figure 1.1 Level 3 integration issues

Integration design

Objectives	Plan of integration, cost/benefit assessment, feasibility, alternatives to integration
Integration specifications	User requirements (intended use, level of training, education), system requirements (hardware, software, computing efficiency)
Decision making	Testing, visualization, ability to replicate integration, decision-support, implementation or advocate alternatives, bidirectional updating and feedback into individual GIS and remote sensing projects

Figure 1.1 (Continued)

increase substantially in number and detail. The relationship between the three levels is a standard hierarchically nested structure; this is where a level 1 component, such as data unity, is divided into a series of level 2 factors, such as information interchange and data availability; and where a level 2 factor such as data availability is divided into level 3 items, such as awareness, publicity, quality, age, etc. (Figure 1.1).

Mesev (1997) only outlines the first three levels (Figure 1.1), but there is no reason why further more refined levels cannot be added. Where schemata have already been documented, for example by Marble (1981) and Davis *et al.*(1991), links between GIS and remote sensing have not been formalized or itemized, and relationships are only presumed. The schema by Mesev (1997) attempts to define the commonest links within a logical structure, and also aims to address direct data coupling, including parallel data acquisition, and analytical operations, with frequent feedback loops and joint decision-making scenarios.

Total integration may be the ultimate goal, yet GIS and remote sensing software have largely retained their independence, even when all technical and methodological issues are sufficiently taken into consideration. For example, there is a conspicuous dearth of literature on total integration in the years since the establishment of the 1990 NCGIA initative. Instead, most studies have tended to focus on the utilization and matching of scale-appropriate thematic information, regardless of source and format (Quattrochi and Goodchild, 1997). Applications spanning both the biophysical and built environments have been facilitated by the expansion in the range of geospatial data, most notably from GPS receivers, and the new breed of remote sensors, such as interferometric synthetic aperture radar (SAR), light detection and ranging (LiDAR), and more recent remote sensors, such as the moderate resolution imaging spectroradiometer (MODIS), the advanced spaceborne thermal emission and reflection radiometer (ASTER), IKONOS and Quickbird.

1.3 Outline of the book

Research and applications throughout this book outline and demonstrate how using data and processing from GIS and remote sensing produces benefits that frequently exceed those from using each technology singularly. Benefits are measured not simply in terms of higher accuracy and greater precision in output, but also on types and levels of information that are otherwise either unavailable or of an inferior quality in one or the other technology.

However, the diverse applications in this book face several common challenges. First, integration can lead to problems of accuracy, uncertainty and scale, which, while affecting any GIS analysis, are often compounded by the integration with remotely sensed data. Chapters 2, 3 and 4 focus almost exclusively on outlining practical solutions for dealing with some of these technical pitfalls. A second major area of concern is the current level of disorganization within GIS and remote sensing technologies. Without a standard method of classifying different operations and data types, it is difficult to develop widely applicable methods of integration. Lack of communication marks a third major obstacle to integration. Chapter 7 notes the need for communication between the remote sensing community and social sciences, while Chapter 9 advocates an exchange of ideas between GIS, remote sensing and the fields of hazard analysis and disaster mitigation. Chapters 6, 8 and 11 showcase the ways in which integration can assist people working in many professions, including urban planning and environmental management. Communication between the academic and professional communities will be an essential factor in the success of integration. Lastly, many of the authors to this book describe their research as a first step towards further integration. They propose better organizational frameworks, more sophisticated applications, and innovative strategies for future interdisciplinary collaboration. Although GIS has long been used to integrate data from various sources, the integration of GIS and remote sensing opens the door to a new world of possibilities.

Chapter 1 attempts to define and conceptualize the rationale, motivation, and expediency behind the integration of data and techniques from the technology of GIS with data and techniques from the technology of remote sensing. It examines whether there is enough scope for overlap and communication and how both technologies have developed concurrently over recent times.

Chapter 2 reiterates the conceptual divisions between GIS and remote sensing and warns of continued *ad hoc* integration if the data integration approach is not replaced by an analysis integration approach based on a taxonomy of system-independent analysis functions. Most existing GIS taxonomies are based on the underlying system and its specific data structure, while various remote sensing systems offer their own unique classification systems. In response, Ehlers proposes an integrated taxonomy based on universal GIS operators and a variety of image processing functions. While somewhat limited, this approach can nevertheless serve as a basis for future progress towards a single, widely applicable, integrated taxonomy for GIS and remote sensing. Another obstacle to total integration is the issue of how to deal with uncertainty. All GIS and remotely sensed data include some level of inaccuracy, but the problem of inaccuracy is compounded when data are transformed from one model of geographic space to another. Ehlers focuses on positional and thematic error, which he identifies as the 'dominant error sources in the integration of GIS and remote sensing'. To support this, an example of a typical GIS/remote sensing analysis (an inventory of land cover over an administrative area) is used to explore positional and thematic uncertainties, along with discussions on line and point errors, confidence regions for line segments, positional uncertainty of boundaries and area objects, and thematic uncertainties of classified remote sensing images. All of these are combined within the 'S-band' model, revealed as a first step towards a more comprehensive model of uncertainty.

Chapter 3 focuses on data fusion, an area of research increasingly prevalent since the inception of 'telegeoprocessing', a term referring to the interaction of GIS, distributed computing systems, telecommunications, GPS, etc. Two of the simplest methods of data fusion, already widely used, are remote sensing output to GIS (e.g. the conversion of a remotely sensed image to a GIS layer) and GIS input to remote sensing interpretation algorithms (e.g. the application of GIS data to remotely sensed images). Simple data fusion is currently being used successfully in commercial urban planning products. However, several fundamental problems must be overcome before more sophisticated techniques become prevalent; for example, the establishment of common standards, the use of compatible legends and scales, and the measurement of the degrees of accuracy. Data fusion, the authors assert, is not possible without first being able to compare data and select the most useful for a given project. Gamba and Dell'Acqua note that it is less important to combine original data than it is to derive useful, comparable information from various sources. By extracting comparable information from different sources, it is possible to view a single type of information from multiple perspectives. The authors provide a round-up of recent approaches to data fusion, such as multi-scale analysis, fuzzy logic and non-parametric and knowledge-based techniques, weighing the pros and cons of each. They then propose a method of integrating GIS and remote sensing into a change detection module, specifically to be used to extract features from a remotely sensed image, analyse change in an existing GIS layer, or detect change using both classification and feature extraction.

Chapter 4 centres on the problems that can occur when integrating data from GIS and remote sensing at different scales, using the 'sampling frame' and the concept of 'support'. The sampling framework, defined as the set of all parameters that determine how data are acquired on a property of interest, affects the scale of spatial variation, present in both raster and vector models; while the support – a term derived from geostatistics and encapsulating the size, geometry and orientation of the space over which an observation is defined – can be thought of as a 'primary scale of measurement'. For instance, variograms and fractal geometry are frequently used to assess the scales of spatial variation in the vector data model, and statistics such as Moran's I and Geary's C can measure spatial autocorrelation – upscaling and downscaling in these allows the size of the support to be altered. Processes can be modelled using spatially distributed dynamic models at appropriate scales; useful when attempting to understand a process better or predict its future behaviour. Atkinson goes on to concentrate on two main types of integration, GIS overlay and remote sensing classification. When combining GIS and remotely sensed data of different scales, degrading the data at finer resolution to match those of the coarser resolution is not always the best choice. It is particularly important to realize that the transformations of data from one form to another impose their own scales. Interpolation techniques, such as IDW and kriging, can be used to transform vector data to raster data, but the smoothing effects of interpolation can produce unwanted consequences. In a discussion of remote sensing land cover classification methods, Atkinson draws attention to problems associated with pixel-based classification, highlighting several advantages of per-parcel classification, soft classification, subpixel allocation and super-resolution mapping. The success of these techniques depends on the scales of measurement, underlying scales of variation, and accurate geometric registration between vector and raster datasets.

Chapter 5 introduces the use of spatial metrics and geostatistics in urban analysis across GIS and remotely sensed data, using techniques such as image interpolation, uncertainty mapping and identification of spatial variability in urban structures. The focus is on land cover and land use, the quintessential dichotomy between biophysical assemblages and anthropogenic exploitation, respectively. Liu and Herold illustrate three empirical studies linking the dichotomy with geostatistics and spatial metrics; the first, classifying images using geostatistics before interpreting the second-order data with spatial metrics; the second, exploring the correlation between population density and urban form; and lastly, reverting to geographically weighted regression to connect urban form and urban growth factors. Overall, these three case studies demonstrate that geostatistics and spatial metrics bring their own strengths and weaknesses to urban analysis.

Chapter 6 illustrates the ways in which GIS and remote sensing can be integrated to reveal spatial characteristics of urban sprawl at the building-unit level. Historically, sprawl research has focused on either demographic-based or physical landscape-based analysis, but concurrent implementation of GIS and remote sensing allows these two branches of investigation to merge. Hasse offers a review of sprawl in the GIS and remote sensing literature, including the variable definitions of sprawl, the concept of smart growth and the analysis of sprawl at the metropolitan and submetropolitan levels. The discussion progresses from simple types of integration (such as land use mapping based on remotely sensed images) to more complicated forms of integration (such as land cover datasets that employ 'land resource impact' indicators). Although geospatial technologies tend to be underused by urban planners and policy makers, Hasse sees great potential for sprawl measurements at the building-unit level, using models that replicate the nested hierarchical structure of urban areas (and may even avoid some of the scale problems mentioned in Chapter 4). The author outlines five geospatial indices of urban sprawl (GIUS) that provide measurements of various forms of sprawl. The five indices are urban density (the amount of land occupied by a housing unit), leap frog (the distance of new housing units to existing housing units), segregated land use (a measurement of land used for similar purposes), highway strip (the amount of land used for strip malls, fast-food restaurants and housing units lining rural highways) and community node inaccessibility (the distance of new housing units to the nearest community centres). The creation of an integrated database could facilitate increasingly sophisticated analyses of building-level urban sprawl.

Chapter 7 reviews a variety of remote sensing applications for urban analysis, but particular emphasis is placed on the estimation of socio-economic information (from remotely sensed images) and the modelling of socio-economic activity (by linking remotely sensed images with GIS data). Various types of socio-economic information can be estimated from remotely sensed images, including population density, employment, gross domestic product and electrical power consumption; and the use of remote sensing allows governments to estimate population in areas where censuses are out of date or unreliable. Population density can be estimated based on types of land use, employment from surface temperature, and GDP and power consumption from nighttime imagery. Furthermore, techniques for population interpolation meld existing population data with additional remote sensing data to create more accurate estimates. Socio-economic indices (e.g. a housing index or quality index) can be created by integrating GIS data with remote sensing data. Wu concludes with a discussion of the advantages and disadvantages of applying remote sensing to urban analysis. Advantages include the frequency with which remotely sensed data can be updated, whilst disadvantages include the lack of dialogue between remote sensing researchers and more traditional social scientists.

Chapter 8 examines the integration of remote sensing, GIS and spatial modelling for sustainable urban planning. It describes historic patterns of urban growth on the outskirts of Atlanta, Georgia, USA, and predicts potential patterns of future development. Using a series of Landsat images of the study area dating from 1973–1999, Yang performs change detection analysis to assess Atlanta's urban expansion, and spatial statistical analysis to identify the forces driving the city's growth. Central to the analyses is the integration of biophysical and socio-economic data at three scales: city, county and census tract. Dynamic spatial modelling is then performed using the SLEUTH urban growth model, with inputs that include remotely sensed and GIS data, such as urban land use, terrain conditions, socio-economic variables and location measures. As a result, the author models two potential scenarios for future urban growth; the first predicts the pattern of urban growth that will occur if current planning strategies remain unchanged and urban sprawl continues unabated, whilst the second scenario predicts the pattern of urban growth that will occur if Atlanta adopts some strategies for 'smart growth' and environmental conservation. This second scenario is favoured because it predicts approximately 50% of the growth that would occur from the first scenario. If geospatial information technology is to be used successfully in sustainable urban planning, Yang asserts that integration is not only desirable, but essential.

Chapter 9 introduces an integrative model for conducting vulnerability analyses – tested on a case study in Los Angeles, California, but remaining portable enough to apply the unique environmental risks and socio-economic context of their study to other places. The authors outline a scenario in which this relationship between the general and the particular is visualized as a hierarchy of nested 'socio-ecological systems'. Specifically, the model integrates GIS and remote sensing data to predict the effects of hypothetical disasters and to highlight locations that are especially at risk. In their case study, susceptible places or 'hot-spots' are identified by a model of urban vulnerability to earthquakes, built on GIS data representing population, building size and geological conditions, as well as remotely sensed imagery used to measure the physical characteristics of the predicted hot-spots. A multiple end-member spectral mixture analysis is used in conjunction with landscape metrics to summarize spatial variation, while census data are used to create an index of wealth for Los Angeles, an index which demonstrates an expected negative correlation between wealth and vulnerability. When constructing their model, Rashed *et al.* borrow techniques from the fields of hazard analysis and disaster management. They argue that future research on GIS and remote sensing integration should be extended beyond the present focus on technological and methodological issues, to include the subject matter and allow its theoretical underlying dynamics to inform the direction of integration.

The last two chapters evaluate the current state of research on environmental applications completed by the mutual interaction of data from GIS and remote sensing. Miller and Rogan in *Chapter 10* focus on biodiversity and ecological representation and analysis, with particular emphasis on species distribution models (SDM) and change detection. In the past, the trend in ecological studies has been to use GIS and remote sensing separately, where GIS functions assist in the calculation of variables pertaining to climate, topography and environmental gradients, while remote sensing contributes information on spectral vegetation indices, structural configuration and land cover classification and change detection. The authors outline how these separate ecological indicators and techniques may be combined within SDM to produce habitat suitability maps, and how levels of biodiversity can be predicted from multiple suitability maps. An early example is the USGS's Gap Analysis Programme, which combines GIS and remotely sensed data to identify potential problems related to biodiversity and species conservation. A more detailed case study by the authors demonstrates an innovative integrative methodology designed to combine five GIS layers (slope, elevation, aspect, vegetation type, and previous fire) with six spectral variables (Kauth Thomas). The methodology harnesses the logic of a hierarchical classification tree with the descriptive and predictive capabilities of generalized linear and generalized additive models to map land cover change in San Diego County, California. However, the reliability and effectiveness of such multivariate predictor models of species distribution can be improved by research that focuses on the extraction of more continuous spectra-based input data, at variable spatial and temporal scales, within more flexible statistical models.

The environmental theme is continued by the last chapter, *Chapter 11* by Shine and Mesev, which centres on the spatial and temporal role of GIS and remote sensing data for monitoring arid-zone ephemeral wetlands. A longitudinal case study from Mauritania in the Sahel region demonstrates how aerial photography, digital topographic maps, GPS readings and satellite sensor data, in combination, can provide valuable information on the location, size, shape and duration of transitional water bodies. The study is in response to the dearth of consistent digital geospatial information on the extent and quality of natural resources in developing countries – and as such modernized databases built on data from GIS and remote sensing are a vital prerequisite for the evolution of sound environmental management policies. In the Mauritania case study, remotely sensed data collected from the 1950s–1980s are used to compare the changes in size of several ephemeral wetlands, along with more detailed information on wetland characteristics from GPS surveys collected during field visits. The authors herald this integrative monitoring strategy as a model of a methodology that can not only help develop sustainable environmental policies in areas affected by ephemeral wetlands but also be applied to many other natural resources in the developing world.

1.4   Conclusions

Total integration has not yet materialized. With the volume of data available and the ease of exchange through the Internet, perhaps the road to full integration is less a computational bottleneck and more a conceptual disparity. As alluded to earlier, and as will be discussed throughout this book, remote sensing is chiefly designed to collect energy-derived geographic information, while GIS is predominantly a data-handling technology capable of comparing, evaluating, modelling and simulating geographic patterns. Conceptually, this difference is almost unsurmountable, and in any case why seek to completely fuse the two technologies into a single system when they seem to function quite satisfactorily side-by-side? If anything, the notion of total integration should refer to attaining a high level of complementary exchange of information and sharing of data processing, rather than some idealized and ultimately unattainable pursuit for homogeneity of data and the relentless strive for identical algorithms.

Another conceptual divergence and major obstacle to integration is how geographic information is represented. Remote sensors record continuous data representing the interaction of energy with the earth, while GIS is predominantly concerned with much more defined and discrete boundaries between geographic features. These two 'views' of reality are conflicting and difficult to operationalize within a single model or system. But these views can be also complementary; they allow the more recent data collected by remote sensors to update and embellish GIS layers, and they allow GIS data to geo-register and help extract information from remotely sensed imagery. Besides, this 'complementary' view is respectful of the fact that most GIS data are derived from remote sensing anyway.

Finally, one further reason for the absence of total integration is expediency. The way many applications 'combine' data from GIS and remote sensing can only best be described as *ad hoc* – not logically and painstakingly within some structured guidelines. Integration to many researchers dealing with geospatial data is *any* process that facilitates the fulfilment of their objectives, regardless of the level of assimilation. So perhaps the definition of integration should remain ambiguous and researchers should instead highlight the strengths of the two individual technologies, rather than strive to attain the redundant and inefficacious goal of absolute amalgamation.

References

Abel, D. J., Kilby, P. J. and Davis, J. R. (1994) The systems integration problem. *International Journal of Geographic Information Systems* **8**, 1–12.

Archibald, P. D. (1987) GIS and remote sensing data integration. *Geocarto International* **3**, 67–73.

de Brouwer, H., Valenzuela, C. R., Valencia, L. M. and Sijmons, K. (1990) Rapid assessment of urban growth using GIS-RS techniques. *ITC Journal* **3**, 233–235.

Brivio, P. A., Colombo, R., Maggi, M. and Tomasoni, R. (2002) Integration of remote sensing data and GIS for accurate mapping of flooded areas. *International Journal of Remote Sensing* **23**, 429–441.

Brown, R. and Fletcher, P. (1994) Satellite images and GIS: making it work. *Mapping Awareness* **8**, 20–22.

Chagarlamundi, P. and Ganulf, J. (1993) Mapping applications for low-cost remote sensing and geographic information systems. *International Journal of Remote Sensing* **14**, 3181–3190.

Davis, F. W., Quattrochi, D. A., Ridd, M. K., Lam, N. S-M., Walsh, S. J., Michaelsen, J. C., Franklin, J., Stow, D. A., Johannsen, C. J. and Johnston, C. A. (1991) Environmental analysis using integrated GIS and remotely-sensed data: some research needs and priorities. *Photogrammetric Engineering and Remote Sensing* **57**, 689–697.

Debinski, D. M., Kindscher, K. M. and Jakubauskas, E. (1999) A remote sensing and GIS-based model of habitats and biodiversity in the Greater Yellowstone Ecosystem. *International Journal of Remote Sensing* **20**, 3281.

Dobson, J. E. (1993) Land cover, land use differences distinct. *GIS World* **6**(2), 20–22.

Dobson, J. E. (1993) Commentary: a conceptual framework for integrating remote sensing, GIS, and geography. *Photogrammetric Engineering and Remote Sensing* **59**, 1491–1496.

Driese, K. L, Reiners, W. A. and Thurston, R. C. (2001) Rule-based integration of remotely-sensed data and GIS for land cover mapping in NE Costa Rica. *Geocarto International* **16**, 35–44.

Ehlers, M., Edwards, G. and Bédard, Y. (1989) Integration of remote sensing with geographic information systems: a necessary evolution. *Photogrammetric Engineering and Remote Sensing* **55**, 1619–1627.

Ehlers, M., Greenlee, D., Smith, T. and Star, T. (1991) Integration of remote sensing and GIS: data and data access. *Photogrammetric Engineering and Remote Sensing* **57**, 669–675.

Faust, N. L., Anderson, W. H. and Star, J. L. (1991) Geographic information systems and remote sensing future computing environment. *Photogrammetric Engineering and Remote Sensing* **57**, 655–668.

Foody, G. M. (1988) Incorporating remotely sensed data into a GIS: the problem of classification evaluation. *Geocarto International* **3**, 13–16.

Gahegan, M. and Ehlers, M. (2000) A framework for the modelling of uncertainty between remote sensing and geographic information systems. *ISPRS Journal of Photogrammetry and Remote Sensing* **55**, 176–188.

Gao, J. (2002) Integration of GPS with remote sensing and GIS: reality and prospect. *Photogrammetric Engineering and Remote Sensing* **68**, 447–453.

Hinton, J. C. (1996) GIS and remote sensing integration for environmental applications. *International Journal of Geographic Information Systems* **10**, 877–890.

Hutchinson, C. F. (1982) Techniques for combining Landsat and ancillary data for digital classification improvement. *Photogrammetry and Remote Sensing* **48**, 123–130.

Janssen, L., Jaarsma, M. and van der Linden, E. (1990) Integrating topographic data with remote sensing for land cover classification. *Photogrammetry and Remote Sensing* **56**, 1503–1506.

Lauer, D. T., Estes, J. E., Jensen, J. R. and Greenlee, D. D. (1991) Institutional issues affecting the integration and use of remotely sensed data and geographic information systems. *Photogrammetry and Remote Sensing* **57**, 647–654.

Marble, D. F. (1981) Some problems in the integration of remote sensing and geographic information systems. Proceedings of the 2nd Australasian Remote Sensing Conference, Canberra, Australia.

Mesev, V. (1997) Remote sensing of urban systems: hierarchical integration with GIS. *Computers, Environment and Urban Systems* **21**, 175–187.

Mesev, V. (1998) The use of census data in image classification. *Photogrammetry and Remote Sensing* **64**, 431–438.

Mesev, V. (2001) Modified maximum likelihood classifications of urban land use: spatial segmentation of prior probabilities. *Geocarto International* **16**, 38–47.

Mesev, V. (ed) (2003) *Remotely Sensed Cities*. Taylor and Francis: London, UK.

Piwowar, J. M., LeDrew, E. F. and Dudycha, D. J. (1990) Integration of spatial data in vector and raster formats in geographical information systems. *International Journal of Geographical Information Systems* **4**, 429–444.

Quattrochi, D. and Goodchild, M. F. (1997) *Scale in Remote Sensing and GIS*. Lewis: New York, NY, USA.

Shi, W. Z., Ehlers, M. and Tempfli, K. (1999) Analytical modelling of positional and thematic uncertainties in the integration of remote sensing and geographical information systems. *Transactions in GIS* **3**, 119–136.

Smits, P. C., Annoni, A., Dellepiane, S. G. (1999) Integration of GIS and remote sensing image analysis techniques. *Proceedings of SPIE* **3871**, 276–283.

Star, J. L., Estes, J. E. and Davis, F. (1991) Improved integration of remote sensing and geographic information systems: a background to NCGIA Initiative 12. *Photogrammetry and Remote Sensing* **57**, 643–645.

Star, J. L., Estes, J. E. and McGwire, K. C. (1997) *Integration of Geographic Information Systems and Remote Sensing*. Cambridge University Press: Cambridge, MA, USA.

Wilkinson, G. G. (1996) A review of current issues in the integration of GIS and remote sensing data. *International Journal of Geographic Information Systems* **10**, 85–101.

Xue, Y., Cracknell, A. P. and Guo, H. D. (2002) Telegeoprocessing: the integration of remote sensing, geographic information systems (GIS), global positioning system (GPS) and telecommunication. *International Journal of Remote Sensing* **23**, 1851–1893.

Zhou, Q. (1989) A method for integrating remote sensing and geographic information systems. *Photogrammetric Engineering and Remote Sensing* **55**, 591–596.

2
Integration taxonomy and uncertainty

Manfred Ehlers

Forschungszentrum für Geoinformatik (FZG) und Fernerkundung, Universität Hochschule Vechta, Germany

2.1 Introduction

Geographic information systems are increasingly seen as an integral part of the modern information and communication society. Improved methods for data access and integration have accelerated this process and scientific advances have paved the way for GIS to be a catalyst for a new evolving discipline of geo-informatics. One of the problems of applying geospatial technology has been the currency, quality, accessibility and completeness of geo-information (GI). Remotely sensed image data, especially from satellites, can be used to generate current, accurate and synoptic information about all parts of the earth as a basis for geoscientific analyses in GIS. Consequently, almost all major GIS software packages now offer at least the possibility to display and query digital images as part of their GIS database. With the advent of the new satellites of 1 m spatial resolution or even better, we will see another push for the integration of remote sensing images into GIS.

The advantages of the integration of GIS and remote sensing have been demonstrated in a large number of application-orientated projects (see e.g. Star *et al.*, 1997). Unfortunately, the merging of remote sensing (and its associated image analysis) and GIS has often resulted in the creation of just another 'dumb' GIS

Integration of GIS and Remote Sensing   Edited by Victor Mesev
© 2007 John Wiley & Sons, Ltd.

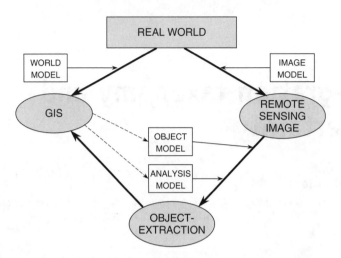

Figure 2.1 Concept for automatic extraction of GIS objects from remote sensing imagery

layer with pictorial information. Integration is restricted to mere georeferencing and image overlay. A complete analysis from a remotely sensed image to a geo-object can be performed only by manual interpretation. GIS and remote sensing information is usually processed separately. The ideal goal should be that GIS objects can be extracted from a remote sensing image to update the GIS database. In return, GIS 'intelligence' (e.g. object and analysis models) should be used to automate this object extraction process (see Figure 2.1). Nevertheless, the current status can still be described primarily as data exchange between a GIS and an image analysis system or an add-on of some image processing functionality to a separate GIS.

Ehlers *et al.*, as early as 1989, presented a concept for a totally integrated system for remote sensing and GIS. They differentiated between three integration levels: (a) two separate systems with a data interface; (b) two principally separate systems with a common user interface; and (c) a totally integrated system (Ehlers *et al.*, 1989). Most of today's GISs offer hybrid processing, i.e. the analysis of raster and vector data. They also have image display capabilities or image analysis add-ons that offer some level (b) functionality (Bill, 1999; Ehlers, 2000). However, geospatial information is usually processed in either raster or vector form and has to be converted into the desired processing or output format. A truly integrated processing option (without prior conversion) does not exist. This is also valid for integrated remote sensing–GIS analyses. The requirements for totally integrated systems are usually defined on an *ad hoc* basis that is driven by project demands or the data sources to be incorporated (Ehlers *et al.*, 1994; Johnston *et al.*, 1997). What is needed is an analysis of the necessary processing components of such an integrated system. The data integration approach has to be replaced by an analysis integration approach. This implies that we need a taxonomy of system-independent analysis functions.

2.2 Taxonomy issues

2.2.1 Taxonomy of GIS operators

If one looks into the functionality of current GIS, it is immediately evident that GIS operations are usually based on the underlying system and its associated data structure. A general description of GIS functions could offer a system-independent view. They are, however, predominantly concerned with low-level functions (see e.g. Laurini and Thompson, 1992; Worboys, 2004). A GIS user, on the other hand, wants to perform a spatial analysis or a comparison of two possible locations for a specific development. He/she is normally faced with a system that offers a huge number of functions that depend upon system and data structures and has to be put together in a specified order to perform the desired analysis. A taxonomy of universal high-level GIS operations that are independent of the system and the data structure is still lacking.

A notable exception is the cartographic modelling (map algebra) approach of Tomlin (1990). However, it is still restricted to raster-based systems. Tomlin structures his cartographic modelling functions into four classes with about 40 subfunctions. These functions are sufficient to perform almost every possible high-level GIS analysis. The strength of his approach is the mathematical rigidity that is incorporated into a computer programming-type GIS language. Structuring the map algebra commands in a procedure allows the composition of very complex GIS analyses. The basic function classes of Tomlin's map algebra are:

- Local functions, e.g. point operations, overlay, recoding.

- Focal functions, e.g. neighbourhood operations, buffering, distance calculation.

- Zonal functions, e.g. attribute operations, intersections.

- Incremental functions, e.g. nearest-neighbour, connectivity, slope, aspect.

Although these functions are system-independent and form the basis of many raster GIS packages in current GIS software, they are still data structure-dependent, i.e. designed for raster GIS. A step further towards a universal GIS language is the approach of Albrecht (1996). Twenty data structure and system-independent high-level GIS functions are grouped into six classes (Table 2.1). The user communicates with the system through a flowchart tool similar to that used in modern GIS and image analysis packages. The difference is that the GIS functions are independent of the underlying system.

Table 2.1   Universal high level GIS operators

Function group	(Sub)functions			
Search	Thematic search	Spatial search	Interpolation	(Re)classification
Location analysis	Buffer	Corridor	Overlay	Thiessen/Voronoi
Terrain analysis	Slope/aspect	Catchment/basins	Drainage/ network	Viewshed analysis
Distribution/ neighbourhood	Cost/diffusion/ spread	Proximity	Nearest neighbour	
Spatial analysis	Pattern/ dispersion	Centrality/ connectivity	Shape	Multivariate analysis
Measurements	Distance	Area		

Modified after Albrecht (1996).

2.2.2   Taxonomy of image analysis operators in remote sensing

Digital image processing started in the early 1970s and is viewed as a young but established discipline. It was influenced by its one-dimensional counterpart, signal processing, by photography and optics, and by the scientific and technological developments in electrical engineering and computer science. Again, its interdisciplinary heritage is clearly visible in the very different descriptions of image processing functionality that can be seen in standard textbooks, such as Pratt (1992) or Sonka *et al.* (1993).

Even in a well-defined application area such as remote sensing, we experience very diverse approaches toward image analysis taxonomies. It is evident that the authors do not want to present a systematic taxonomy of image analysis functions. Nevertheless, textbooks mix hardware, sensors, systems and operations or present structures that are inconsistent with a rational image processing taxonomy (Ehlers, 2000). This inconsistency when dealing with image analysis taxonomies is an impediment for the development of a stronger theoretical background for the design and implementation of integrated GIS. Without such a theoretical basis, the only way to GIS–remote sensing integration seems to be a project-driven *ad hoc* approach with limited usefulness and applicability.

2.2.3   An integrated taxonomy

To set up a taxonomy of data structure and system-independent GIS–image analysis functions, one has to start from either the remote sensing or the GIS side. The 20 universal operators from Table 2.1 currently represent the only taxonomy that meets the requirements stated above. Although the grouping can be debated (are 'buffer' and 'corridor' really different functions, or does 'interpolation' belong to

Table 2.2 Universal image processing functions for integrated GIS

Function group	(Sub)functions			
Preprocessing	Parametric radiometric sensor correction	Parametric geometric sensor correction		
Geometric registration	Deterministic techniques	Statistical techniques (interpolation)	Automated techniques (matching)	Error assessment
3D image analysis	Ortho-image generation	DEM extraction		
Atmospheric correction	Deterministic approaches	Histogram-based manipulations (point operations)	Filtering	Image enhancement
Feature/object extraction	Unsupervised techniques	Supervised techniques	Model-based techniques	Error assessment

After Ehlers (2000).

the 'search' or to the 'spatial analysis' group?), the operators can be used as starting points for an iterative approach. Based on typical remote sensing analyses, four groups with 17 image-processing functions were selected to be added to the 20 universal GIS operators (Table 2.2). The derivation of these functions is a first step and is based on an in-depth analysis of remote sensing literature and intensive project experiences (Richards and Jia, 1999; Ehlers and Schiewe, 1999; Ehlers, 2000; Schiewe and Ehlers, 2003; Jensen, 2005).

It has to be noted, however, that the operators presented in Tables 2.1 and 2.2 are not sufficient to define and describe the complete functionality of integrated GIS. Still required is a thorough analysis of hybrid processing capabilities, i.e. functions that allow a joint analysis of remote sensing and GIS information. Still to be investigated how polymorphic techniques can be used to extend the capacities of the universal high-level GIS/image processing functions. The operator 'overlay', for example, should be able to process image–image, GIS–image and GIS–GIS overlays without a different name for every function option. First results of such polymorphisms were investigated, for example, by Jung (2004). Additional functions have to be developed that extend the capabilities of integrated GIS beyond the sum of the single components. Three-dimensional (3D) urban information systems created from GIS and remote sensing can be seen as an example of these extensions.

2.3   Uncertainty issues

2.3.1   Uncertainty in geographic information

The advantages of an integrated geoprocessing framework have been confirmed by many examples, yet it is also evident that the issue of accuracy and errors within this integration process has to be addressed. Good science requires statements of accuracy by which the reliability of results can be understood and communicated. When accuracy is known objectively, it can be expressed as error; when it is not the term 'uncertainty' applies (Hunter and Goodchild, 1993). Thus, uncertainty covers a broader range of doubt or inconsistency and, in the context of this chapter, includes error as a component. The understanding of uncertainty as it exists in geographic data remains a problem that is only partly solved (see e.g. Story and Congleton, 1986; Goodchild and Gopal, 1989; Veregin, 1995; Ruiz, 1997; Worboys, 1998; Gahegan and Ehlers, 2000; Zhang and Goodchild, 2002). Without quantification, the reliability of any results produced remains problematic to assess and difficult to communicate to the user. GISs provide a whole series of tools with which data can be manipulated without offering any control over misuse. To that instance, Openshaw *et al.* (1991) state:

> A GIS gives the user complete freedom to combine, overlay and analyse data from many different sources, regardless of scale, accuracy, resolution and quality of the original map documents and without any regard for the accuracy characteristics of the data themselves.

This is a serious issue; without quantification of uncertainty, the results themselves may only be considered as qualitative information, and this greatly devalues their merit in both a scientific and a practical sense. To compound the problem, in the fusion of activities from remote sensing and GIS, an integrated approach to managing geographic information is required. This must necessarily support many different types of data (Ehlers *et al.*, 1991), gathered according to different models of geographic space (Goodchild, 1998), each possessing different types of inherent errors and uncertainties (Chrisman, 1991). As well as providing individual support for these different models of space, it is necessary to explicitly include methods that keep track of uncertainty as data are changed from low-level forms (such as remotely sensed image data) to the higher-level abstractions required by digital cartography and GIS (such as objects and themes).

Whether a particular dataset can be considered suitable for a given task depends on many different criteria and, despite the fact that various aspects of uncertainty can be measured objectively, their importance will be largely determined by the task. The overall goal when modelling uncertainty is therefore threefold: (a) to produce a statement of uncertainty to be associated with each dataset, so that an objective statement of reliability may be reported; (b) to develop methods to propagate uncertainty as the data are processed and transformed; and (c) to ultimately determine the suitability of a dataset for a given task ('fitness for use'). Another goal is to communicate uncertainty information to the user (e.g. Hunter and Goodchild, 1996).

A useful framework for handling uncertainty, recognizing the separate error components of value, space, time, consistency and completeness, was proposed as early as 1978 by Sinton and later embellished by Chrisman (1991). Uncertainty in geographic data can be described in a variety of alternative ways; such as those provided by Bédard (1987), Miller *et al.* (1989) and Veregin (1989). Although different, these approaches all have a number of aspects in common, including the observation that uncertainty itself occurs at different levels of abstraction. For example, positional and temporal error describe uncertainty in a metric sense within a spatiotemporal framework, whereas completeness and consistency represent more abstract concepts describing coverage and reliability, and are consequently more problematic to describe.

Uncertainty in its many forms has been on the research agenda of the GIS and remote sensing community for at least two decades, gaining much of its early momentum from the very first research initiative of the US National Centre for Geographic Information and Analysis (NCGIA; Goodchild and Gopal, 1989). Work to date on uncertainty addresses the inherent errors present within specific types of data structures (e.g. raster or vector) or data models (e.g. field or object). The effects of error propagation and analysis within these various paradigms have been studied by Veregin (1989, 1995), Openshaw *et al.* (1991), Goodchild *et al.* (1992), Heuvelink and Burrough (1993), Ehlers and Shi (1997), Leung and Yan (1998), Shi (1998), Arbia *et al.* (1999), Zhang and Kirby (2000), Zhang and Stuart (2001) and Shi *et al.* (2003). In a recent compendium on uncertainty in geographic information, Zhang and Goodchild (2002) investigated methods for uncertainty assessment for continuous variables (fields), categorical variables (classes) and objects. Despite the progress made to date, they concluded that:

> . . . academics, technologists, government information agencies, the general public and the commercial sector must work together to take advantage of the benefits of geographical information in new applications, while being fully informed of the nature and implications of the associated uncertainties. Scientists and workers lead the leap forward.

This does not sound like a problem solved.

2.3.2 Uncertainty in the integration of GIS and remote sensing

Even if we restrict uncertainty description to one specific problem, the integration of remote sensing and GIS, a generally applicable solution does not exist. In

1990, the error analysis research group of the NCGIA Initiative 12 (integration of remote sensing and GIS) identified the research on uncertainty as one of the major challenges in the integration of these two technologies (Lunetta *et al.*, 1991). Remote sensing scientists have always had the need to quantify errors that were associated with the processing of remotely sensed data. Most efforts have gone into the error analysis of rectification and registration processes and of information extraction or multispectral classification techniques (see e.g. Ehlers, 1997; Gongalton and Green, 1999)

The problem of modelling uncertainty as the data are transformed through different models of geographic space was addressed by Lunetta *et al.* (1991), Gahegan (1996), and Gahegan and Ehlers (2000). A typical path taken by data captured by satellite, then abstracted into a suitable form for GIS, is shown in Figure 2.2, and involves four models. Continuously varying fields are quantified by the remote sensing device into image form, then classified and finally transformed into discrete mapping objects. The overall object extraction process is sometimes referred to as 'semantic abstraction' (Waterfeld and Schek, 1992), due to the increasing semantic content of the data as it is manipulated into forms that are easier for people to work with.

When transforming data between different conceptual models of geographic space, the uncertainty characteristics in the data may change, in that techniques used to transform the data also alter the inherent uncertainty. In addition, these techniques may introduce further uncertainties of their own. Furthermore, many of the abstraction techniques employ combined data with different uncertainty characteristics. Consequently, two interrelated problems must be addressed:

- How do the uncertainty characteristics of data change as data are transformed between models?

- How do the transformation methods used affect and combine the uncertainty present in the data?

One of the consequences of the traditional separation of GIS and remote sensing activities into distinct communities and separate software environments is that there is an artificial barrier between the two disciplines. Therefore, the integration of these

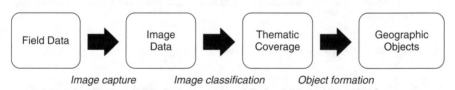

Figure capture   Image classification   Object formation

Figure 2.2 Continuum of abstraction from field model to object model. After Gahegan and Ehlers (2000)

two branches of science is to some extent an artificial problem. As a result, there is no easy flow of metadata between systems. Interoperability is often restricted to the exchange of image files or object geometry and the problem of managing uncertainty is compounded. The four stages shown in Figure 2.2 represent the four models of geographic space, namely field, image, theme and object (or feature) models, and are typical of models in the integration of GIS and remote sensing activities. These models only represent the conceptual properties of the data and can be considered here as independent from any particular data structure that might be used to encode and organize the data.

Gahegan and Ehlers (2000) developed an integrated error simulation model for the transition from field (raw remote sensing) data to geo-objects. The description of uncertainty followed that proposed by Sinton (1978). It covers the sources of error as they occur in remote sensing and GIS integration (although other approaches may be equally valid). Uncertainty is restricted to the following properties: (a) value (including measurement and label errors); (b) spatial; (c) temporal; (d) consistency; and (e) completeness. Of these, measurement and label errors, as well as uncertainties in space and time, can apply either individually to a single datum or to any set of data. The latter two properties of consistency and completeness can only apply to a defined dataset, since they are comparative (either internally amongst data or to some external framework). The findings of Gahegan and Ehlers (2000) are summarized in Table 2.3. Other uncertainty issues in the integration of the two spatial technologies can be related to scale and representation of the data (Bruegger, 1995; Guptill and Morrison, 1995) or the provision of lineage information (Lanter, 1991).

New research on uncertainty deals with the development of advanced processing techniques for information extraction from remotely sensed images. The inclusion of contextual information (textures, neighbourhood) and object- or segment-based analysis techniques, together with the application of fuzzy set theory and artificial intelligence, challenge the standard image-processing strategies (Wang, 1993; Ryherd and Woodcock, 1996; Lucieer and Stein, 2002; Ibrahim *et al.*, 2005). In a special issue of the *International Journal of Remote Sensing* on 'Uncertainties in Integrated Remote Sensing and GIS', the editors conclude:

> Within the framework of uncertainties in integrated remote sensing and GIS, we can describe the uncertainties in terms of positional accuracy, attribute and thematic accuracy, temporal accuracy, logical consistency, and completeness. In this special issue, we mainly address the modeling of uncertainty in terms of attribute and positional accuracy. Relatively less attention is paid to the issue of completeness or temporal uncertainties. Modeling uncertainties in newly emerging datasets, such as laser scanning data, high-resolution satellite images, InSar, and high spectral satellite images, will be an area for future research (Shi *et al.*, 2005).

Table 2.3 Types of uncertainty and their sources in four models of geographic space

	Field	Image	Thematic	Object
α	Measurement error and precision	Quantization of value in terms of spectral bands and dynamic range	Labelling uncertainty (classification error)	Identity error (incorrect assignment of object type), object definition uncertainty
β	Locational error and precision	Registration error, sampling precision	Combination effects when data represented by different spatial properties are combined	Object shape error, topological inconsistency, 'split and merge' errors
χ	Temporal error and precision	(Temporal error and precision are usually negligible for image data)	Combination effects when data representing different times is combined	Combination effects when data representing different times is combined
δ	Samples/readings collected or measured in an identical manner	Image is captured identically for each pixel, but medium between satellite and ground is not consistent; inconsistent sensing, light fall-off; shadows	Classifier strategies are usually consistent in their treatment of a dataset	Methods for object formation may be consistent, but often are not. Depends on extraction strategy
ε	Sampling strategy covers space, time and attribute domains adequately	Image is complete, but parts of ground may be obscured (clouds, trees)	Completeness depends on the classification strategy (is all the dataset classified or are only some classes extracted?)	Depends on extraction strategy. Spatial and topological inconsistencies may arise as a result of object formation

α, data or value; β, space; χ, time; δ, consistency; ε, completeness.
From Gahegan and Ehlers (2000).

2.4   Modelling positional and thematic error in the integration of remote sensing and GIS

Positional and thematic uncertainties seem to be the dominant error sources in the integration of GIS and remote sensing. We present an analytical model for the combination of these uncertainties, based on the work of Shi *et al.* (1999).

2.4.1   Positional and thematic uncertainties

A typical integrated GIS–remote sensing analysis might require an inventory of the land cover over a certain administrative area. The boundary of the area was digitized from a map, and is available in GIS format. The land cover types are obtained from a classified remote sensing image, using a typical maximum likelihood (ML) classification (see Figure 2.3). For a thorough analysis, not only do the areas of the different land cover classes have to be calculated but also their respective errors (e.g. the spatial distribution of an error or uncertainty parameter). In this example, two types of spatial data are involved: GIS and classified remote sensing data. It can now be assumed that the GIS data have only positional uncertainties resulting from the digitization and measurement process. The classified remote sensing data,

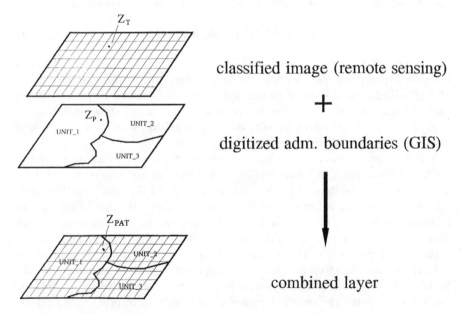

Figure 2.3   Land resources inventory using remote sensing and GIS techniques. Z_T is a pixel in the classified image; Z_P is a GIS point close to a boundary; Z_{PAT} is a point in the combined layer with the same location as Z_P and Z_T

on the other hand, contain only thematic uncertainties, which are the result of a statistical classification process.

2.4.2 Problem formulation

On a digitized map in a GIS, we know $P[Z_P(\mathbf{X}) \in O_j]$, i.e. the probability that point Z_P belongs to a certain area O_j (e.g. a county). This is referred to as the positional uncertainty indicator. $\mathbf{X}$ is a vector in two-dimensional Euclidean space, normally $\mathbf{X} = (X, Y)^t$, which describes the geometric location of point Z_P. Typically, a remote sensing image is classified using a maximum likelihood (ML) classification. For a given pixel $Z_T(\mathbf{X})$ that is geometrically located at $\mathbf{X}$, its thematic characteristics are determined by its position in n-dimensional feature space, where n is the number of spectral bands of the remotely sensed image. Using ML classification techniques, the probability that this pixel belongs to a specific class C_i, i.e. $P[Z_T(\mathbf{X}) \in C_i]$, can be calculated (Richards and Jia, 1999). C_i is one class type of the whole class category set. This set is usually predefined in a supervised ML classification procedure. The probability value per class can be used as a thematic uncertainty (or certainty) indicator. Other accuracy estimators and other classification techniques exist, which means that any other value for $P[Z_p(\mathbf{X})]$ can be used in the thematic uncertainty description. After combining the classified remotely sensed image and the GIS boundary layer, the combined probability can now be estimated:

$$P(Z_T(\mathbf{X}) \in C_i) \wedge (Z_p(\mathbf{X}) \in O_j) \tag{2.1}$$

For this combination, Ehlers and Shi (1997) developed an analytical model that was based on: (a) the positional uncertainty of an area object; (b) the thematic uncertainty of a classified remote sensing image; and (c) the combination of positional and thematic uncertainties. To model the uncertainties of a two-dimensional object (e.g. an area feature) in a vector GIS, they distinguished two regions, the (fuzzy) boundary region and the interior region (Figure 2.4).

The difference between interior and boundary regions is based on positional uncertainty. An object in a vector-based GIS is built of line segments. The error at the end points of these line segments (or vertices of the area object) directly affects the positional uncertainty of the object boundary. The region affected by boundary errors is called the fuzzy boundary region. The interior region of an area object is the region where the effects of the positional uncertainties of its vertices are below a certain threshold (e.g. outside of 3σ for a Normal distribution) and can therefore be ignored. The interior region is influenced mainly by thematic uncertainties that originate from errors in the classification process. Therefore, its spatial uncertainty distribution can be derived by probability methods commonly used in remote sensing classification. In the fuzzy boundary region, however, both positional and thematic uncertainty factors contribute to the overall uncertainty.

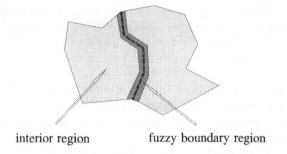

interior region   fuzzy boundary region

Figure 2.4 Fuzzy boundary and interior regions. The positional uncertainties of boundary points affect only the boundary region and not the interior region. The uncertainty of the interior region is mainly determined by the thematic uncertainty

2.4.3   Modelling positional uncertainty

2.4.3.1   Line errors

Positional errors of GIS objects are mainly related to the problem of deciding, for each object, which boundaries best describe its geometric extension. The identification of the object boundary is uncertain, usually due to errors in the digitization and/or measurement procedures. This geometric approach works best for those cases where the boundaries have been identified. In such cases, a decision has been made regarding which areas belong to a specific object. In photo-interpretation, for example, the decision about the extension of the object is often separated from the measurement of the geometry of the object boundaries, because these are often digitized after the interpretation was performed. In surveying, boundary identification and measurement are generally combined in one process, but the fact that these are still two operations remains. In such typical cases for the creation of GIS databases, it might be that boundary curvatures can only be approximately represented by means of discretized description of chains of points and line segments. Line errors depend on the sampling density and can be derived from the analysis of digitized points.

Point errors have been intensively investigated in disciplines such as geodesy and surveying. For positional errors of lines, Perkal (1956, 1966) developed the epsilon band model, which was further applied by Chrisman (1982) and Blakemore (1984). The epsilon band is constructed as a simple buffer of constant width (epsilon) on either side of a measured line, and the true location of the line is assumed to be contained within the epsilon band. However, there is no provision to describe the distribution of a measured line segment around its true location. Zhang and Tulip (1990) and Caspary and Scheuring (1992) derived variances in X and Y direction for an arbitrary point on the line segment, based on the laws of error propagation. Dutton (1992) and Caspary and Scheuring (1992) used Monte Carlo simulation techniques to model the distribution of line segments and other geometric features. The simulation approach, however, cannot provide a general analytical form of the solution. To combine positional and thematic uncertainties, it is necessary to derive the spatial structures and their error distributions for geometric features (Ehlers and Shi, 1997).

To model the combined uncertainty, the positional uncertainties of area objects must first be addressed. The basic geometric element of a GIS object is the point. Two connected end points compose a line segment. A line feature is composed of line segments. An area object is defined by its boundary line features. Thus, we have a hierarchical procedure for building area objects: points, line segments, line features, boundary line features and area objects. As mentioned above, there exist well-developed models for the description of positional uncertainty of points in geodesy and surveying. These models can be applied to solve the problem of uncertainty description of fuzzy boundary regions.

A point is geometrically described by its coordinates. Consequently, errors in coordinates constitute one of the components of positional uncertainty in a GIS. The second component is caused by sampling and approximation of curved line features by a sequence of straight line segments. This error is directly associated with line segments. The coordinates of a point in a GIS are usually the result of measurements, and of various processing steps, so each operation involved can add to the overall error. These errors can be classified into three groups: blunders, systematic errors and random errors. As techniques exist for the detection of systematic errors and blunders, the uncertainty model deals only with random errors. If the final coordinates can be expressed as a function of the original measurements, the error characteristics of a GIS point can be analytically determined by applying the laws of error propagation. Assuming that the errors are Normally distributed and not correlated, standard statistical techniques can be applied.

With these assumptions, line and area boundary errors can be analytically constructed based on Normal distribution of point errors and a Normal error distribution perpendicular to the direction of line segments. Figure 2.5 demonstrates the probability distribution around the true location of a line with end points ζ_1 and ζ_2 (for more details, see Shi, 1994; Ehlers and Shi, 1997).

2.4.3.2   *Confidence region for line segments*

As the true location of a GIS point is usually not known, the model has to be modified to estimate from the error model the confidence region of measured points and the edges that they connect. Let Z_1 and Z_2 be the measured locations of the true end points ζ_1 and ζ_2. The next step is to derive a confidence region for the line segment $Z_1 Z_2$ because the precise location of the true line $\zeta_1 \zeta_2$ is not known. This procedure is based on a sampling density for line digitizing that allows the construction of a line based on interpolated lines from point measurements.

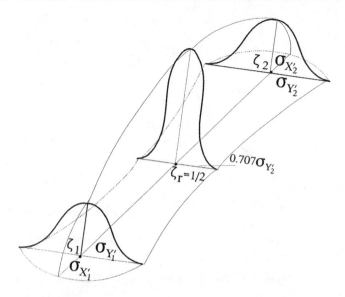

Figure 2.5 Probability distribution around the true locations $\zeta_1\zeta_2$

The derivation of the confidence region J_r is based on the distribution of the points on the line. If the variance matrix of an arbitrary point Z_r on the line segment is known, we can derive χ^2 distributed statistics of X_r and Y_r, and then the confidence region J_r, for any arbitrary point ζ_r on the line between ζ_1 and ζ_2. J_r is constructed so that it contains ζ_r, which has a predefined confidence level γ, while all other ζ of the line segment are contained in their respective confidence regions. This involves an upper boundary condition, leading to the inequality (Ehlers and Shi, 1997):

$$P(\zeta_r \in J_r, r \in [0, 1]) > \gamma \qquad (2.2)$$

The confidence region J of a line segment is the union of the sets 0_r for all $r \in [0,1]$. One region, J_r, is a set of points $(x, y)^t$, satisfying:

$$X_r - c \leq x \leq X_r + c$$
$$Y_r - d \leq y \leq Y_r + d \qquad (2.3)$$

where:

$$c = k^{1/2}\{[(1 - r)^2 + r^2]\sigma_{xx}\}^{1/2}$$
$$d = k^{1/2}\{[(1 - r)^2 + r^2]\sigma_{yy}\}^{1/2} \qquad (2.4)$$

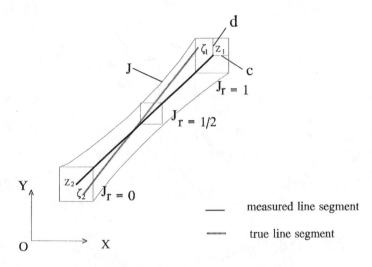

Figure 2.6   Confidence region of a line segment

The parameter k depends on the selected confidence level γ and can be obtained from a χ^2 table, $k = \kappa^2_{2;(1+\gamma)/2}$. For example, for $\gamma = 0.90$, $(1 + \gamma)/2 = 0.95$, we have $k = 5.99$. A detailed derivation can be found in Shi (1994).

It is easy to verify that the maximum value of $\{[(1 - r)^2 + r^2]\}^{1/2}$ is obtained for $r = 0$ or 1, whereas the minimum value is at $r = \frac{1}{2}$. This means that the confidence region is smallest at the centre of the line segment and largest at the end points (Figure 2.6). With a risk expressed by the confidence level, we can state that the true line is somewhere inside the confidence region.

2.4.3.3   Positional uncertainty of boundary line features

A boundary is composed of one or (usually) several line segments. In describing positional uncertainties of boundary line features, two problems need to be solved: the confidence region of boundary line features and its probability distribution. The confidence region of a line feature can be constructed by the union of the confidence regions of the constituent line segments. It provides an uncertainty zone of the spatial extension of a line feature.

In describing positional error distributions of boundaries, one of the major problems is to understand the nature of the uncertainty in the region where two line segments join (see Figure 2.7). Within this region, we have two different probability values that a given point Q belongs to an object A: one is based on the uncertainty distribution of line segment L_1, which is denoted as $P_1(Q \in A)$. The other is based on the uncertainty distribution of line segment L_2, i.e. $P_2(Q \in A)$. To obtain the overall probability that point Q belongs to object A, we need the combined

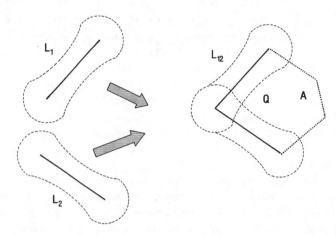

Figure 2.7   The uncertainty of a line feature L_{12} is determined by the uncertainties of line segments L_1 and L_2 and by considering the uncertainty of points in the joint region, e.g. point Q. The line feature L_{12} is part of the boundary of polygon A

uncertainty distribution $P_1 \wedge_2 (Q \in A)$ of the line feature L_{12}, which is composed of L_1 and L_2.

If $P_1(Q \in A)$ and $P_2(Q \in A)$ were independent, we could directly use the product law of probability theory to solve this problem, i.e. $P_1 \wedge_2 (Q \in A) = P_1(Q \in A)P_2(Q \in A)$. Because the two line segments share one end point, however, we cannot assume independence. An alternative is the use of fuzzy set theory (Zadeh, 1978). To apply fuzzy set theory, we need to treat the probability values as corresponding membership values. For example, the probability that Q belongs to object A is treated as the membership value that element Q belongs to a fuzzy set A. We can follow this approach because the subjective interpretation of probability considers probability as a measure of belief. Thus, we can state (Shi, 1994):

$$P_{1 \wedge 2}(Q \in A) = \min[P_1(Q \in A), P_2(Q \in A)] \qquad (2.5)$$

which means we can use the minimum operation between the probabilities for segments L_1 and L_2 within the joint region of two line segments. Accordingly, we can calculate the uncertainty value for the composed line features of a polygon boundary.

2.4.3.4   Positional uncertainty of area objects

An area object is defined as an area enclosed by a boundary line feature. The positional uncertainty of an area object is determined by that of the boundary. As shown above, the positional uncertainty affects mainly the fuzzy boundary region. The positional uncertainty of an area object is described by the probability that a point (x, y) belongs to the area object (O), i.e. $P[(x, y) \in O] \in [0, 1]$. When a point 'moves' from the outside to the interior region of the area object, the probability changes gradually from 0 to 1. The probability distribution of the boundary region is determined by the cumulative probability function perpendicular to the boundary. Therefore, it allows a continuous range to describe uncertainty for area objects. In comparison, the so-called epsilon band model to describe the 'point-in-polygon' problem, i.e. the uncertainty of an area object enclosed by a polygon, can only distinguish five relationships between a point and the area object (Figure 2.8). These are: 'definitely in' (point 5), 'definitely out' (point 1), 'possibly in' (point 4), 'possibly out' (point 2) and 'ambiguous' (point 3).

Using the probability distribution of line segments, we can describe the relationship between a point and an area object by probability values varying continuously within [0,1] (see Figure 2.8). This approach provides a quantitative indicator of uncertainty and, moreover, facilitates the combination with thematic uncertainty indicators. We can also characterize the positional uncertainty of an area object by computing a probability frequency distribution. For example, with 10 probability interval classes (i.e. 0–0.1, 0.1–0.2, ... , 0.9–1.0), we can calculate the result of Table 2.4 for the area object in Figure 2.8. Of a total of 1638 pixels, 649 (about 40% of the area) have a probability of less than 90% that they belong to the area object. The rate (40%) is dependent on the error of the vertices and the size of the area object. If the error of the vertices is relatively small compared to the size of the area object, the rate will be significantly lower than 40%.

2.4.4   Thematic uncertainties of a classified image

Thematic uncertainty in this context refers to the thematic uncertainty inherent in a classification derived from a remote sensing image. The reason for this uncertainty is that the classification is based on limited evidence. For demonstration purposes,

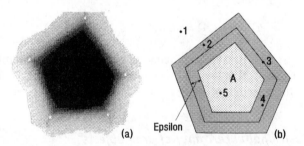

Figure 2.8 'Point-in-polygon' description of uncertainty for an area object. For this developed uncertainty model (a), the uncertainty values vary continuously within [0,1]; for the epsilon band model (b), uncertainty is distinguished only in five qualitative levels. Adapted from Ehlers, M. and Shi, W. Z. (1997) Error modelling for integrated GIS. *Cartographica* **33**, 11–21, courtesy of the University of Michigan Press

Table 2.4 Frequency distribution of probability values of Figure 2.8

Prob.	0–0.1	0.1–0.2	0.2–0.3	0.3–0.4	0.4–0.5	0.5–0.6	0.6–0.7	0.7–0.8	0.8–0.9	0.9–1.0	Total
Area	0	0	0	0	20	107	154	170	198	989	1638

The row 'Prob.' shows the intervals of probability values; 'Area' is the number of pixels located within a certain interval (e.g. 20 pixels have a probability value between 0.4 and 0.5); 'Total' is the total number of pixels in the study area. From this table, one can see the positional uncertainty of the area object indicated by the numbers of pixels located within each probability interval.

we will make use of the probability vectors in the well-developed ML classification technique as a basic thematic uncertainty indicator. The parameters used for classification in this technique are estimated from training samples, and then a probability vector is calculated for each pixel in the image defining the likelihood of specific class membership. The pixel is then assigned to the class with the maximum probability (Richards and Jia, 1999). For example, in an image with four classes (urban, water, forest, agriculture), a pixel with the probability vector:

$$\{P[Z_P(\mathbf{X}) \in \text{urban}] = 0.35, P[Z_P(\mathbf{X}) \in \text{forest}] = 0.32, P[Z_P(\mathbf{X}) \in \text{agriculture}] = 0.30,$$

$$P[Z_P(\mathbf{X}) \in \text{water}] = 0.03\}$$

will be assigned to the class 'urban'. The other probability values are usually ignored. In the above case, however, there is only weak evidence that this pixel actually belongs to the class urban (the probability is only 35%). If the maximum probability value for each pixel is retained, the certainty of the classification result can be described. If we attach the probability value $P[Z_P(\mathbf{X}) \in \text{urban}] = 0.35$ to the classification result, it is easy to see that this classification is very uncertain. If the whole probability vector could be attached, a user might further learn that the pixel could just as well be forest or agriculture (with probabilities of 32% and 30%, respectively). Based on the techniques discussed above, we can now combine positional and thematic uncertainty assessments in the integration of GIS and remote sensing data.

2.4.5 Modelling the combined positional and thematic uncertainties

The 'S-band' model was developed to combine positional and thematic uncertainties (Shi 1994; Shi and Ehlers, 1996). There are two alternatives within the S-band model; one is based on the product rule, the other is based on a certainty factor model with probabilistic interpretation. If two data layers are from two different data sources, for example one is from GIS and another is from remote sensing data, they are independent from each other. Thus, the product-rule-based approach can be used to combine positional and thematic uncertainties. The uncertainty values are within the range [0,1]. For the general case with non-zero correlation between the data layers, a procedure was developed based on a certainty factor model with probabilistic interpretation. This model is also used in expert system design for uncertainty-based reasoning. With this model, the range of uncertainty expressions is extended from [0,1] to [−1,1]. This is particularly important for a reasoning that includes uncertainty indicators covering both positive and negative ranges.

When combining a GIS layer with positional error and remote sensing data, it can be assumed that the uncertainties of the two datasets are independent of each other. We can therefore directly apply the product-rule model to calculate the combined positional and thematic (PAT) uncertainty (see Figure 2.9):

$$P[Z_T(\mathbf{X}) \in C_i] \wedge [Z_p(\mathbf{X}) \in O_j] = P[Z_t(\mathbf{X}) \in C_i]P[Z_p(\mathbf{X}) \in O_j] \qquad (2.6)$$

where $P[Z_T(\mathbf{X}) \in C_i]$ is the probability that $Z_T(\mathbf{X})$ belongs to class C_i and $P[Z_p(\mathbf{X}) \in 0]$ is the probability that point $Z_p(\mathbf{X})$ belongs to area object O_j.

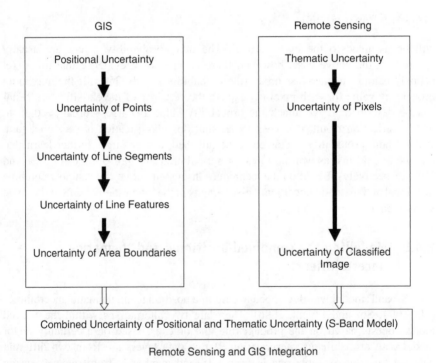

Figure 2.9 Diagram of modelling PAT uncertainty of objects using the S-band model. After Gahegan and Ehlers (1997)

To demonstrate the effects of thematic uncertainties, a Landsat test image with four classes (urban, forest, grassland, water) was classified using ML classification techniques. To compute the frequency distribution of probability values, we again use the 10 intervals (0.0–0.1), (0.1–0.2) . . . (0.9–1.0) (Table 2.5).

Given a boundary layer with positional uncertainty indicators and a classified remote sensing image with thematic uncertainty indicators, an overlay operation in GIS can be used to solve the uncertainty combination problem. The combined uncertainty for the test area is given in Table 2.6 and combines information about the size of each land cover class within the area and the uncertainty that is associated with this class. For example, the size of the land cover type 'forest' within the test area is 1280 pixels, of which 80 pixels have a certainty between 0.1 and 0.2, 344 pixels between 0.2 and 0.3, and 240 pixels between 0.9 and 1.0.

One result that is evident from a comparison of Tables 2.5 and 2.6 is that after the PAT combination more pixels have lower probability values. For example, within the interval 0.9–1.0, the number of pixels for the class 'urban' is reduced from 13

Table 2.5 Maximum likelihood classification and thematic uncertainty, expressed as frequency distribution of maximum probability values

Prob.	0.0–0.1	0.1–0.2	0.2–0.3	0.3–0.4	0.4–0.5	0.5–0.6	0.6–0.7	0.7–0.8	0.8–0.9	0.9–1.0	Sum
Urban	0	0	56	13	3	9	7	5	11	13	117
Grassland	0	0	18	17	4	15	7	11	12	138	222
Water	0	0	12	0	0	0	0	0	0	7	19
Forest	0	0	394	57	65	48	65	73	126	452	1280
Total											1638

The column 'Sum' shows the total area in pixels that were classified as 'urban', 'grassland', etc. The table also shows the distribution of each class within the probability intervals, thus describing the uncertainty of the classification.

Table 2.6 The statistics of a classified image with combined uncertainty indicators

Prob.	0.0–0.1	0.1–0.2	0.2–0.3	0.3–0.4	0.4–0.5	0.5–0.6	0.6–0.7	0.7–0.8	0.8–0.9	0.9–1.0	Sum
Urban	0	24	42	7	7	15	12	4	2	4	117
Grass	0	9	19	13	18	26	25	29	29	57	222
Water	0	4	8	0	2	4	1	0	0	0	19
Forest	0	80	344	62	95	93	109	133	124	249	1280
Total:											1638

The column 'Sum' is the total area in pixels that were classified as a certain land cover class. The table also describes the PAT uncertainty of each land cover class by providing the number of pixels within each probability interval.

to 4, for 'grassland' from 138 to 57, for 'water' from 7 to 0 and for 'forest' from 452 to 240. On the other hand, the number of pixels with high uncertainty values has increased. For example, within the interval 0.1–0.2, the number of pixels for the class 'urban' has increased from 0 to 24, for 'grassland' from 0 to 9, for 'water' from 0 to 4 and for 'forest' from 0 to 80. With the combined uncertainty, we can get a quantitative description of the extent to which the overall uncertainty is increased by the combination of positional and thematic uncertainties.

2.5 Conclusions

In this chapter, we have presented an overview of taxonomy and uncertainty issues for the integration of GIS and remote sensing. It is evident that progress has been made over the last 20 years or so. It is also clear that for both problems research is still required. In particular, an integrated taxonomy and system-independent description of high-level GIS and remote sensing operators is still missing. Very scant research has been done on this issue.

This is different from the issue of uncertainty descriptors for the integrated GIS–remote sensing datasets. The importance of this issue is evidenced by a number of research initiatives and scientific publications. It can be seen from the S-band model that the quantification of uncertainties for the integration of remote sensing and GIS is possible. It is also evident that within the other aspects of uncertainties, e.g. time, topology and completeness, this model is just a first step towards a complete uncertainty model – if it ever exists.

References

Albrecht, J. (1996) Universal Analytical GIS Operations. PhD Thesis, ISPA-Mitteilungen 23, University of Vechta, Germany.

Arbia, G., Griffith, D. and Haining, R. (1999) Error propagation modelling in raster GIS: adding and rationing operations. *Cartography and Geographic Information Science* **26**, 297–315.

Bédard, Y. (1987) Uncertainties in land information databases. Proceedings of Auto-Carto 8, Baltimore, MD, USA, 175–184.

Bill, R. (1999) GIS-Produkte am Markt – Stand und Entwicklungstendenzen. *Zeitschrift für Vermessungswesen* **6**, 195–199.

Blakemore, M. (1984) Generalization and error in spatial data bases. *Cartographica* **21**, 131–139.

Bruegger, B. P. (1995) Theory for the integration of scale and representation formats: major concepts and practical implications. In Frank, A. U. and Kuhn W. (eds), *Spatial Information Theory*. Lecture Notes in Computer Science No. 988. Springer: Berlin, 297–310.

Caspary, W. and Scheuring, R. (1992) Error-band as measures of geographic accuracy. Proceedings of the European GIS 1992, Munich, Germany, 226–233.

Chrisman, N. R. (1982) A theory of cartographic error and its measurement in digital data bases. Proceedings of Auto-Carto 5, Alexandria, VA, USA, 159–168.

Chrisman, N. R. (1991) The error component in spatial data. In Maguire, D. J., Goodchild, M. F. and Rhind D. W. (eds). *Geographical Information Systems*, Vol. 1. Longman: Harlow, UK, 165–174.

Congalton, R. G. and Green, K. (1999) *Assessing the Accuracy of Remotely Sensed Data: Principles and Practices*. Lewis: Boca Raton, FL, USA.

Dutton, G. (1992) Handling positional uncertainty in spatial databases. Proceedings of the 5th International Symposium on Spatial Data Handling, Charleston, SC, USA, 460–469.

Ehlers, M. (1997) Rectification and registration. In Star J. L., Estes J. E. and McGwire K. C. (eds), *Integration of Remote Sensing and GIS*. Cambridge University Press: New York, NY, USA, 13–36.

Ehlers, M. (2000) Integrated GIS – from data integration to integrated analysis. *Surveying World* **9**, 30–33.

Ehlers, M. and Schiewe J. (eds). (1999) *Geoinformatik 99: Ausgewählte Themen der Forschungsgruppe GIS/Fernerkundung*. Materialien Umweltwissenschaften Vechta, 5, University of Vechta, Germany.

Ehlers, M. and Shi, W. Z. (1997) Error modelling for integrated GIS. *Cartographica* **33**, 11–21.

Ehlers, M., Edwards, G. and Bédard, Y. (1989) Integration of remote sensing with GIS: a necessary evolution. *Photogrammetric Engineering and Remote Sensing* **55**, 1619–1627.

Ehlers, M., Greenlee, D. D., Star, J. L. and Smith, T. R. (1991) Integration of remote sensing and GIS: data and data access. *Photogrammetric Engineering and Remote Sensing* **57**, 669–675.

Ehlers, M., Steiner, D. R. and Johnston, J. B. (eds). (1994) *Requirements for Integrated Geographic Information Systems*. Environmental Research Institute of Michigan, Ann Arbor, MI, USA.

Gahegan, M. N. (1996) Specifying the transformations within and between geographic data models. *Transactions in GIS* **1**, 137–152.

Gahegan, M. and Ehlers, M. (2000) A framework for modelling of uncertainty in an integrated geographic information system. *ISPRS Journal of Photogrammetry and Remote Sensing* **55**, 176–188.

Goodchild, M. F. (1998) Different data sources and diverse data structures: metadata and other solutions. In Longley, P. A., Brooks, S. M., McDonnell, R. and Macmillan W. (eds), *Geocomputation: A Primer*. Wiley: London, UK, 61–74.

Goodchild, M. F. and Gopal, S. (eds). (1989) *The Accuracy of Spatial Databases*. Taylor and Francis: London, UK.

Goodchild, M. F., Guoqing S. and Shiren, Y. (1992) Development and test of an error model for categorical data. *International Journal of Geographical Information Systems* **6**, 87–104.

Guptill, S. C. and Morrison, J. L. (1995) *Elements of Spatial Data Quality*. Elsevier Science: New York, NY, USA.

Heuvelink, G. B. M. and. Burrough, P. A. (1993) Error propagation in cartographic modelling using Boolean logic and continuous classification. *International Journal of Geographical Information Systems* **7**, 231–246.

Hunter, G. and M. F. Goodchild. (1993) Mapping uncertainty in spatial databases, putting theory into practice. *Journal of Urban and Regional Information Systems Association* **5**, 55–62.

Hunter, G. and Goodchild, M. F. (1996) Communicating uncertainty in spatial databases. *Transactions in GIS* **1**, 13–24.

Ibrahim, M. A., Arora M. K. and Ghosh S. K. (2005) Estimating and accommodating uncertainty through the soft classification of remote sensing data. *International Journal of Remote Sensing* **26**, 2995–3007.

Jensen, J. R. (2005) *Introductory Digital Image Processing*, 2nd edn. Prentice Hall: Upper Saddle River, NJ, USA.

Johnston, J. B., Ehlers, M. Steiner, D. R. and Gomarasca, M. A. (eds). (1997) *New Developments in Geographic Information Systems*. Environmental Research Institute of Michigan, Ann Arbor, MI.

Jung, S. (2004) *HYBRIS: Hybride räumliche Analyse Methoden als Grundlage für ein integriertes GIS*. PhD Thesis, University of Vechta, Germany (CD Publication).

Lanter, D. P. (1991) Design of a lineage-based meta-data base for GIS. *Cartography and Geographic Information Systems* **18**, 255–261.

Laurini, R. and Thompson D. (1992) *Fundamentals of Spatial Information Systems*. Academic Press: New York, NY, USA.

Leung, Y. and Yan, J. (1998) A locational error model for spatial features. *International Journal of Geographical Information Science* **12**: 607–620.

Lucieer, A. and Stein A. (2002) Existential uncertainty of spatial objects segmented from satellite sensor imagery. *IEEE Transactions on Geoscience and Remote Sensing* **40**, 2518–2521.

Lunetta R. S., Congalton, R. G., Fenstermaker, L. K., Jensen, J. R., McGwire K. C. and Tinney, L. R. (1991) Remote sensing and geographic information system data integration: error sources and research issues. *Photogrammetric Engineering and Remote Sensing*, **57**, 677–687.

Miller, R., Karimi, H. and Feuchtwanger, M. (1989) Uncertainty and its management in geographical information systems. Proceedings of CISM '89, Canadian Institute of Surveying and Mapping, Ottawa, Canada, 252–259.

Openshaw, S., Charlton M., and Carver, S. (1991) Error propagation: a Monte Carlo simulation. In Masser, I. and Blakemore M. (eds), *Handling Geographic Information*. Longman: Harlow, UK, 78–101.

Perkal, J. (1956) On epsilon length. *Bulletin de l'Academie Polonaise des Sciences* **4**, 399–403.

Perkal, J. (1966) On the length of empirical curves. Discussion Paper 10, Michigan Inter-University Community of Mathematical Cartographers, Ann Arbor, MI, USA.

Pratt, W. K. (1992) *Digital Image Processing*, 3rd edn. Wiley: New York, NY, USA.

Richards, J. A. and Xiuping, J. (1999) *Remote Sensing Digital Image Analysis*. Springer: Berlin.

Ruiz, M. O. (1997) A causal analysis of viewshed error. *Transactions in GIS* **2**, 85–94.

Ryherd, S. and Woodcock, C. (1996) Combining spectral and textural data in the segmentation of remotely sensed images. *Photogrammetric Engineering and Remote Sensing* **62**, 181–194.

Schiewe, J. and Ehlers, M. (eds). (2003) *Geoinformatik 03: Ausgewählte Themen der Forschungsgruppe GIS/Fernerkundung*. Materialien Umweltwissenschaften Vechta (MUWV), 17, Vechta, Germany.

Shi, W. Z. (1994) Modelling positional and thematic uncertainties in the integration of remote sensing and geographic information systems. PhD Thesis, ITC Publication 22, Enschede, The Netherlands.

Shi, W. Z. (1998) A generic statistical approach for modelling errors of geometric features in GIS. *International Journal of Geographical Information Science* **12**, 131–143.

Shi, W. Z. and Ehlers, M. (1996) Determining uncertainties and their propagation in dynamic change detection based on classified remotely sensed images. *International Journal of Remote Sensing* **17**, 2729–2741.

Shi, W. Z., Cheung, C. K. and Zhu, C. Q. (2003) Modelling error propagation in vector-based GIS. *International Journal of Geographical Information Science* **17**, 251–271.

Shi, W. Z., Ehlers, M. and Molenaar, M. (eds). (2005) Uncertainties in integrated remote sensing and GIS. *International Journal of Remote Sensing* **26**, 2909–3120.

Shi. W., Ehlers M. and Tempfli, K. (1999) Analytical modelling of positional and thematic uncertainties in the integration of remote sensing and geographical information systems. *Transactions in GIS*, **3**: 119–136.

Sinton, D. (1978) The inherent structure of information as a constraint to analysis: mapped thematic data as a case study. In Dutton, G. (ed.), *Harvard Papers on Geographic Information Systems* 6. Addison-Wesley: Reading, MA, USA.

Sonka, M., Hlavac, V. and Boyle, R. (1993) *Image Processing, Analysis and Machine Vision.* Chapman and Hall: London.

Star, J. L., Estes, J. E. and McGwire, K. C. (eds). (1997) *Integration of Remote Sensing and GIS.* Cambridge University Press: New York, NY, USA.

Story, M. and Congleton, R. G. (1986) Accuracy assessment: a user's perspective. *Photogrammetric Engineering and Remote Sensing* **52**, 397–399.

Tomlin, D. (1990) *GIS and Cartographic Modelling.* Prentice-Hall: Englewood Cliffs, NJ, USA.

Veregin, H. (1989) Error modelling for the map overlay operation. In Goodchild, M. F. and Gopal, S. (eds), *Accuracy of Spatial Databases.* Taylor and Francis: London, UK, 3–19.

Veregin, H. (1995) Developing and testing of an error propagation model for GIS overlay operations. *International Journal of Geographical Information Systems* **9**, 595–619.

Wang, F. (1993) A Knowledge-Based Vision System for Detecting Land Changes at Urban Fringes, *IEEE Transactions on Geoscience and Remote Sensing*, **31**, 136–145.

Waterfeld, W. and Schek, H. J. (1992) The DASBDS Geokernel – an extensible database system for GIS. In Turner, A. K. (ed.), *Three-Dimensional Modelling with Geoscientific Information Systems.* Kluwer Academic: Amsterdam, The Netherlands, 45–55.

Worboys, M. F. (1998) Computation with imprecise geospatial data. *Computers, Environment and Urban Systems* **22**, 85–106.

Worboys, M. F. (2004) *GIS: A Computing Perspective*, 2nd edn. Taylor and Francis: Keele, UK.

Zadeh, L. A. (1978) Fuzzy sets as a basis for a theory of possibility. *Fuzzy Sets and Systems* **1**, 3–28.

Zhang, G. Y. and Tulip, J. (1990) An algorithm for the avoidance of sliver polygons and clusters of points in spatial overlay. Proceedings of the 4th International Symposium on Spatial Data Handling, Zurich, Switzerland, 141–150.

Zhang, J. and Goodchild, M. F. (2002) *Uncertainties in Geographical Information.* Taylor and Francis: London, UK.

Zhang, J. and Kirby, R. P. (2000) A geostatistical approach to modelling positional errors in vector data. *Transactions in GIS* **4**, 145–159.

Zhang, J. and Stuart, N. (2001) Fuzzy methods for categorical mapping with image-based land cover data. *International Journal of Geographical Information Science* **15**, 175–195.

3

Data fusion related to GIS and remote sensing

Paolo Gamba and Fabio Dell'Acqua

Telecommunications and Remote Sensing Laboratory, Universita degli Studi di Pavia, Italy

3.1 Introduction

The integration of geographic information systems and remote sensing essentially involves combining data provided by both, sensibly and consistently. In this chapter we examine the various aspects of data fusion issues related to GIS–remote sensing integration. In the first section we explore the reasons why such fusion is desirable or, in some cases, badly needed. In the second section, the issues and problems inherent in GIS and remote sensing data fusion are introduced, while in the third, possible solutions are outlined. A final section draws some conclusions.

3.2 Why do we need GIS–remote sensing fusion?

Spatial information technologies are increasingly used in combination to support a variety of activities related to, or based on, spatial phenomena. These technologies are many and diverse, but can be roughly grouped into the following areas:

- Remote sensing and photogrammetry.
- Computer vision and artificial intelligence.
- Virtual reality and multimedia.

Integration of GIS and Remote Sensing   Edited by Victor Mesev
© 2007 John Wiley & Sons, Ltd.

It is widely recognized that all technologies contribute to the completion of many varied applications. Their integration often follows a scheme in which the data manager, represented by the GIS, has a pivotal role. All of the other systems are linked to the data manager, activated and accessed through mutual data interaction. When it comes to applications, remote sensing data clearly have the firmest connection with physical reality. As such, a very tight connection between these data and the underlying GIS must be ensured for effective implementation. This connection has various aspects, which are separately treated in the following subsections.

In a more general framework, we justify the necessity of GIS–remote sensing data fusion by looking at the new frontiers of interaction among GIS, distributed computing systems, telecommunications networks, multimedia processing and global navigation systems (GPS). This comprehensive framework, called *telegeoprocessing* (Xue *et al.*, 2002) is being gradually introduced into our lives, driven by many market forces and science developments. First of all, the possibility of sharing data seamlessly on various platforms by using different communication tools is one of the driving forces of modern life, as demonstrated by the exponential growth of cellular phones and GNS receivers for road navigation. From the standpoint of remotely sensed data, the expansion of transmission bandwidths and computational power of on-board processors has led to huge image databases and the processing of very large files. Moreover, web interfaces (and the rise of WebGIS systems) are now commonly implemented, along with the exploitation of grid computing techniques currently tested for earth observation data levels 2 and 3 processing. Similarly, real-time and on-board processing of remotely sensed data requires that many sources are jointly considered, such as meteorological data for atmospheric corrections and digital terrain models for geometric corrections. Often, these data are stored in vector format and in GIS layers that need to be quickly and precisely combined with raw imagery to provide user-friendly products at the right time and immediately to the market.

Consequently, the need for data fusion involving GIS and remote sensing is already high and steadily increasing to the point where a taxonomy is necessary. Ideas on how to deal with such a complex scenario are proposed and discussed in the following sections, in relation to 'application-driven' examples outlining the juxtaposition of remotely sensed and vector data into a GIS (as in de la Ville *et al.*, 2002) and how to exploit GIS and remotely sensed data sources.

3.2.1 Remote sensing output to GIS

The simplest way to integrate remote sensing with a GIS is to convert interpreted remotely sensed data into a layer just like any other in a GIS. This is standard practice for classification schemes where the thematic map is deliberately disassembled into layers, each one representing a particular class. Another remote sensing to GIS example is to extract geometric features from algorithms, such as those devoted to building or man-made structure identification and recognition, into vectors that can be translated into GIS formats.

Examples of these applications include the work by Chica-Olmo *et al.* (2002) and Mason *et al.* (1997). Both presented decision systems using a GIS and both used remotely sensed data, or maps derived from remote sensing data, as inputs to the decision processes. However, such situations are far from a trivial use of remotely sensed data within a GIS because they require the selection of the most relevant subsets of information. In Chica-Olmo *et al.*, whose study was devoted to gold-rich area identification, remotely sensed imagery and existing thematic maps were inserted in the GIS database, while in Mason *et al.* the target was informal settlement management. 3D data representing shelters and shacks were extracted from aerial surveys and combined with other information to understand the dynamics of rapidly changing suburban and deprived areas. In all of these examples, we need to stress that no data fusion techniques are applied as long as remote sensing output is just translated into GIS. However, often this step is just a part of a methodology with broader aims, devoted to using the new layer to perform a spatial analysis integrating the already available information in the GIS with the new one, independently extracted from remotely sensed imagery. In this sense, the operations mentioned above may be seen as preliminary to data fusion and, since they allow further processing, as a fundamental and mandatory part of the fusion process.

3.2.2 GIS input to remote sensing interpretation algorithms

There is a second easy option for fusion of GIS and remote sensing, i.e. the exploitation of GIS data as known input to supervised image interpretation algorithms. GIS layers are considered as *a priori* knowledge, useful for understanding the remotely sensed scene. Therefore, they may be used for defining training areas accurately and efficiently, reducing the costs of ground surveys or the errors due to manual image interpretation. GIS data, indeed, are always considered as more precise, at least from the point of view of labels and legends, while their spatial accuracy relative to remotely sensed imagery may be insufficient. This approach is applied in Walter (2004) to object extraction and identification. In that study GIS was used to define training areas for a first classification. By considering a number of object properties computed in the image using GIS object boundaries as input, GIS spatial information was also used in reference to object identification. These properties (e.g. mean grey and textural measures values) were used to assign new objects to already existing classes by a second step, such as an object supervised classification.

A similar approach, applied to the task of road database verification, was proposed in Bonnefon *et al.* (2002). It is interesting to note that this paper discriminates between the two tasks of GIS data verification and GIS data update. For the first task, remote sensing interpretation is guided by *a priori* known information coming from GIS layers. In the second step, by contrast, remote sensing leads the process, and GIS may be used to infer the identification of an object (in this work a linear feature) by using the local spatially adjacent information, e.g. the land cover of the surrounding areas. Another important difference with respect to the previous work is that GIS- and remote sensing-based information (i.e. the output of a combination of line and edge detectors) are fused, using Dempster–Shafer rules to validate the matching between the detected features and those already in the database.

Generally speaking, GIS information is used in many research applications to introduce ancillary information that it is not possible to capture from remote sensors. This is true, for instance, for land use classes (as opposed to land cover) that are recognized using existing and digitized maps or administrative boundaries (Tapiador and Casanova, 2002). Similarly, GIS information is used to integrate geomorphological maps with representative soil profiles for landfill identification and classification (Maksud Kamal and Midorikawa, 2004). In any situation the geographical information system is used to manage remotely sensed data in conjunction with sparser or very different data sources, and to apply procedures fully exploiting the spatial distribution of all the sources. In other words, the geographical nature of the sources is dealt with in the GIS, while the spectral (point-wise) or local (in a local window) information is classified or extracted in the remote sensing processing environment.

3.2.3   Example: urban planning check and update

As an example of GIS–remote sensing data fusion at this level, we can examine checks and updating in urban planning. Understanding the content of a remotely sensed image may be the basis for identifying the elements of the observed urban area to cross-check the development with plans and assess land use efficiency and other parameters of interest. This is a crucial point in urban area management and is related to urban planning, urban planning and urban environmental monitoring. A number of techniques are available for this task, and many of them refer to GIS as the final output stage of the procedure. A list of satellite-based techniques related to this task (mainly based on very high-resolution sensors in the optical and near-infrared bands) is presented in Guindon (1997). A selection of techniques relevant to urban planning and based on the fusion of GIS and remote sensing information follows.

Approaches to understanding images of urban areas differ. One is to attempt full-scene interpretation (not necessary in our case), another is to aim at a predefined goal (which can be, for example, detecting buildings) and yet another is to simply aid the human interpreter in accomplishing his/her task. Given the limited scope of the images to be analysed and the defined purpose of the analysis, our choice of systems will mostly focus on goal-directed systems, without disregarding other types of systems where appropriate. Xiuwan (2002) provides a comparison of many methods, rather than a single method. Post-classification is used, and the importance of ancillary data (possibly integrated into a GIS) and improving single-date classification performance are stressed. As in other post-classification methods, emphasis is placed on improving single-date classification performance. The post-classification analysis is equivalent to adding GIS information after the spectral analysis is completed, i.e. feeding a GIS system some rules with the output of one or more urban classifier. In this work the classifiers are statistical and pixel-based, while spatial relationships are uncovered and exploited in the GIS framework.

Finally, an example of knowledge-based integration of GIS data into a remotely sensed data spectral classifier is found in Stefanov et al. (2001), where ASTER data have been used to analyse a large urban area in Arizona. Moreover, they were coordinated with many different layers of information. The results are encouraging, and have been recently proposed in a second paper on the same area using LANDSAT data, which proves the robustness of the system. A very interesting point of view may be added to this series of work if a change detection chain is considered for urban map updating. In this case, region-based change detection is usually required by the final user, therefore a cost function, taking into account the user requirements, is often the key to successful acceptance of the final change map. This consideration leads us quite naturally to the topic of the integration of GIS and remote sensing data. An interface to and from a GIS layer is usually essential for providing information that is valuable for final users, especially in urban areas. This may lead to a direct comparison of one-date classification to a GIS layer, as in Prol-Ledesma et al. (2002), or drive the classification by means of the already considered GIS layer (Janssen and Molenaar, 1995; Smith and Fuller, 2001).

In summary, the problem of urban planning through urban area monitoring and exploiting remotely sensed images and GIS has been faced using many approaches. Indeed, there are techniques that use remote sensing for GIS analyses, or GIS layers as ancillary data for remote sensing classification. The cited methods have been widely used and analysed and are thus robust and mature enough to be used and integrated into marketable products. Many expert systems based on available GIS products or procedures built around remote sensing commercial classifiers are already operative and attempt, as we will see in the following section, to translate GIS–remote sensing data fusion into common practice.

3.3 Problems in GIS–remote sensing data fusion

This section highlights some of the problems involved in the fusion of remotely sensed data and GIS features or layers. These may be summarized by the following:

- Both the GIS and the remotely sensed data are delivered by companies and public agencies which do not agree on a standard. Even if, in principle, an urban plan definition update system works for any GIS and any image, it should be carefully tuned for the different datasets. This, in turn, reduces its range of applicability and generalization properties.

- There is also a difference between the legends of the two information sources. It is pretty clear, for example, that the concept of 'street' in a GIS municipal database is different from 'street' in a topological sense, which is (at best) what it is possible to extract from remotely sensed data. The question of how to translate between administrative boundaries and geographical boundaries is another well-known example.

- It is also interesting to investigate how to combine related information from very different sources, such as the street network, street widths, and real-time traffic measurements from *in situ* sensors. In order to feed traffic models and forecast performance of traffic control systems, this combination is necessary.

As a result of the previous points, we may assume that we need algorithms for information rather than data fusion. In other words, even if the algorithms highlighted above may be labelled in a generic way as belonging to data fusion methods, information fusion is mandatory to achieve a true integration of the two sources. In the urban mapping update example, the information coming from remotely sensed data is geometrical and related to the physics of the streets as patches of particular land covers. The information from GIS may be related to administrative, traffic and business value information. The use of both data types may improve knowledge of the urban environment, seen as a complex system, with many inputs and outputs. In the following we would like to give a more detailed analysis of each of these points.

3.3.1   Lack of consistent standards

As noted above, the first problem is related to the lack of a unique standard, not only for GIS but also for remote sensing imagery. This problem is continuously exacerbated by the common practice of assigning each new satellite (if not sensor) its own data format, requiring new input routines to insert them in existing software.

On the GIS side, even if many national and international bodies (see European Commission, 2004; Rao *et al.*, 2002) are working toward the definition of a common standard for GIS, there are at the moment many possible ways to store data in GIS. These ways often reflect proprietary techniques by GIS vendors, and are disclosed to the public only for inter-operability issues. Finally, as pointed out in many documents, it is considerably difficult to promote environmental monitoring relying solely on national data sources. Therefore the need for harmonization at an international level is really important.

To this end, the European Union recently established an initiative to create an infrastructure for spatial information in Europe (INSPIRE). It will help to make geographical information more accessible through harmonization efforts in the fields of spatial data specifications, interoperability of spatial data services and data-sharing policies. In this way INSPIRE will support a wide range of purposes, directly or indirectly related to environmental policies and sustainable development (INSPIRE, 2005). Of course, this is just an attempt, and related to European regional issues only, so it does not provide a final answer to the question of when a common standard will be available. However, it looks as though INSPIRE, at least theoretically, marks a step in the right direction. The hope in the scientific community is that a general standard will finally arise. Unfortunately, this will happen only when the market requires it. As a matter of fact, the limited number of actors in the GIS–remote sensing arena, most of them public (i.e. able to impose their own standards) or very large companies, has reduced the possibility of such a solution for now. The greater availability of data and the need for solution may provoke a development in a different direction in the near future.

3.3.2   Inconsistency of GIS–remote sensing accuracy, legends and scales

The need for data integration also requires a common base. This is especially different for remotely sensed data and GIS layers, since they do not come from the same source and therefore carry different uncertainties, and refer to different legends and work on different scales.

As for uncertainties, please refer to Chapter 2 of this book for a detailed analysis. It is sufficient to say in this analysis that it is mandatory not only to take care of the uncertainties of the multiple sources and track the combined uncertainties in order to manage them. It is also important to consider the nature of such uncertainty, to weight the importance of the sources. For instance, spectral uncertainties at the sensor translate into lower affordability of the classification results and lower mapping accuracy, but this is true at the full resolution of the data, while at a coarser scale this may not necessarily be the case. As a result, the uncertainty of the fusion between remotely sensed data and a GIS data layer is different with respect to the scale of the fusion.

In another example, the positional accuracy of a very high-resolution satellite image affects the results of road extraction procedures if the inputs include the shape of the object as it was originally stored in a GIS layer. In this case, the relative accuracy is far more important than the absolute accuracy.

As for legends, we may use as an example urban area characterization and problems related to land use (to which urban planners and their GIS usually refer) and land cover (which is what remote sensors are able to classify). The algorithms used to obtain classify land cover have long been tested, and for many datasets and sensors it is clearly evident to the scientific and user communities what it is possible

Table 3.1 Urban land use classes from European and US classification systems (level 3)

CORINE land cover (EU)	USGS modified Anderson scheme (US)
1.1.1 Residential continuous urban fabric	2.11 Single-family residential
1.1.2 Residential discontinuous urban fabric	2.12 Multi-family residential
1.2.1 Industrial or commercial, public and private unit	2.21 Commercial/light industry
1.2.2 Road and rail networks and associate land	2.22 Heavy industry
1.2.3 Port areas	2.23 Communications and utilities
1.2.4 Airports	2.25 Agricultural business
1.3.1 Mineral extraction sites	2.26 Transportation
1.3.2 Dump sites	2.27 Entertainment/recreational
1.3.3 Construction sites	
1.4.1 Green urban areas	
1.4.2 Sport and leisure facilities	

to obtain and at which accuracy level. On the other hand, land use legends usually refer to economic or administrative information, and cannot be directly related to the physics of the interaction between electromagnetic waves and the earth surface. As a result, it is fairly difficult to match patterns of land cover into land use classes if one starts from remotely sensed imagery. This becomes almost impossible for some objects whose characterization relies on truly independent sources. Take, for example, the difference between post offices and shops, which is important from the user's point of view but is negligible or null from the point of view of remote sensing.

The view is further complicated by the huge difference between legends proposed and used for the same task by different agencies/administrations. In Table 3.1 we offer a very basic comparison between level 3 urban legends adopted by the European Union and the USA. Many classes clearly overlap, but in general no direct translation is possible without a segmentation and reclustering procedure. No GIS layer for urban areas is at the moment truly exploitable across the Atlantic Ocean.

Finally, the scale issue is also a very important one, since the same object assumes a different meaning with respect to the scale and its neighbourhood. The analysis of a geographic scene at different resolutions reveals different details and information, and often leads to very different results. The ability, or the possibility, to combine information referring to different scales is one of the points discussed in the current literature and implemented in recent image interpretation algorithms and software. For example, e-cognition software (Benz *et al.*, 2004) is able to refine the original segmentation of the scene based on spectral and local spatial information (e.g. texture), using a hierarchical network of relationships among the objects into which the scene was segmented at different scales. The same approach, although the segmentation and the scale analysis is integrated into one single step, is used in Hall *et al.* (2004). A watershed segmentation is applied to the scene after it is decomposed in a continuous set of scale-dependent features, each one with its own particular range of significance.

3.3.3   Different nature of the two sources

The difference between the two sources is among the most pre-eminent reasons for a reduced diffusion of fusion approaches. It may be paired with the problem of defining the 'quality' of data coming from these sources for a given application, before even starting to think of any fusion.

Indeed, the differences prevent the comparison of the sources, which is the first step to an efficient data fusion. How can we manage to fuse datasets if we do not know their utility for a given task? Simply using all the available data sources usually does not help, because an informed choice of the 'best' sources often gives the same results and without the need to consider very different sources. Sometimes, this process to reduce the inputs (usually labelled a feature-reduction process) to the most meaningful ones results in a better solution of the application problem. If the number of features to be used is very large, it may be that, using all of them, a worse result is obtained. This is the well-known Hughes phenomenon (Hughes, 1968), which appears, for instance, when dealing with very fine spectral (e.g. hyperspectral) remotely sensed data (Shahshahani and Landgrebe, 1994), where the discrimination capability of the different land covers is somehow masked by the large or very large number of wavelengths used to characterize the scene.

Feature reduction, however, calls for the definition of a 'discriminatory' index. This, in turn, requires an assessment of the utility of each dataset (or part of a dataset) for the specific task. The task deserves careful attention, as shown by the survey among stakeholders reported in Meeks and Dasguota (2004). The geospatial information utility developed as a result, even if proposed by the authors as a proof-of-concept more than a real-world application, paves the way for more refined solutions. The point is that the user and producer communities must agree on a number of indicators that are useful to all of them to understand the really important characteristics of a GIS or a remotely sensed dataset. A statistical analysis of the weight to be used for the combination of these indicators may be run consequently to reduce the available dataset to the most useful for a given application or task.

The above-cited research work is based on very simple and basic indicators, such as currency, vertical and horizontal accuracies, datum and form of the data. However, the issue of indicator definition for many applications is currently the subject of extensive analysis. Generally speaking, the definition of these indicators, connected to a given task for which remotely sensed and GIS data may be exploited, may be considered as the first step toward a full realization of this idea.

We may, for instance, refer to urban indicators. A standard definition of an urban indicator is a measure that summarizes information about a particular subject. It provides a reasonable response to specific needs and questions asked by decision makers and policy makers. It also provides an objective description of the conditions of the urban area as they relate to the goals of the community. Indicators reflect the trend of development and also provide quantitative and qualitative information. Based on the previous sections, we may state that urban remote sensing may be useful to urban indicators in the following areas:

• *Biophysical/health*, i.e. air and water quality, food quality, land contamination, public sanitation, etc.

• *Biophysical/use*, i.e. conservation of natural resources, open space, etc.

• *Infrastructure*, i.e. access to basic services (health, sanitation facilities, garbage disposal structures, etc.), energy, transportation (road networks, railroads, bridges), city composition (residential, industrial areas).

More complex applications correlate remotely sensed data with other indicators, in areas such as:

• *Demographics*, i.e. population characteristics (density, distribution), age structure, fertility, family structure, etc.

• *Socio-economic*, i.e. income, political stability, urban governance, financial performance, unemployment, literacy, etc.

Consequently, there are still serious problems when implementing urban indicators in the policy cycle. Most major economic aggregates that measure the health of the urban economy, such as city product, investment or trade, are not detectable by remote sensing. Data that describe the condition of the population, infrastructure and the environment may be partially available, depending on the topic, but are seldom collected within a consistent framework. Data that measure the internal spatial structure of the city, its economy and the distribution of opportunities are the only data that offer a sufficient degree of precision.

Moreover, since few final users understand the potential and limitations of urban remote sensing, there is often no clear idea which territorial indicators related to urban analysis may be extracted from remote sensing data. Finally, due to technical characteristics of the algorithms applied to the collected data, indicators derived from remote sensing sources tend to emphasize physical environmental data and land cover rather than land use. As a matter of fact, they are often computed without the involvement of the final users and with a limited use of ancillary sources.

GIS can help overcome some of these problems. Access to information, the ability to process it, and the generation of alternative outcomes are essential in supporting decision making. By maintaining databases of a wide variety of spatial and non-spatial data, GIS plays a fundamental role in the planning process. Information integration on the basis of a shared geographic footprint is seen as one of the major strengths of GIS. Beyond data integration and access facilitation, the power of GIS as a decision support tool stems from its analytical and synthesizing potential.

As a final note, we would like to stress that urban indicators are currently used inside the Global Monitoring for Environment and Security initiative (GMES) promoted by the European Union and the European Space Agency. The GMES Urban Services (GUS; Gamba and Dell'Acqua, 2004) project has been investigating some of these indicators and the availability of remote sensing (mainly) and GIS data useful with respect to them.

3.3.4   Need for information rather than data fusion

All the previous examples illustrate that information is more important than original data fusion. In other words, since it is difficult to combine the original data because of the above-mentioned problems, it may be interesting to extract some kind of common information and combine the two views of the same thing. The usual example for a digital cadastral is the extraction of the shape of a built-up structure, recognized as a building, using remotely sensed data and its combination/comparison with existing shape files already available in the GIS system.

Following the definition of data fusion in Wald (1998), and considering the interesting discussion in Varma *et al.* (2003), we may say that remote sensing–GIS data fusion may be classified as a data compilation/assimilation problem in one of its more complex forms. Information in different formats (raster and vector), with different scales and with heterogeneous spatial and spectral attributes, should be merged, also taking into account different legends and quality measures.

The major problem in this combination is the inability to measure, or the complexity of measuring *a priori*, the quantity of information carried by a data source. In turn, this prevents a wise selection of the best subset for a given task. Even using advanced statistical approaches, such as evidential reasoning, that allow the combination of different sources in a common framework, there is no way to define the degree of evidence of the datasets. Therefore, users are forced to design optimization rules based on algorithmic performances, rather than really understanding the relevance of each source to the application being tested. For instance, in Peddel and Ferguson (2002) the suggested solutions to this problem are the exhaustive search in the multidimensional input space, computationally very demanding, or the 'independent search,' which assumes that each input variable is unrelated to the others.

3.3.5 Example: population mapping through remote sensing

As an example of the problems inherent to remote sensing–GIS fusion, we consider population mapping. The application is very interesting, especially for fast-growing or third-world countries, where urbanization creates huge changes in the population distribution in the country. It is also important for humanitarian purposes, for monitoring informal settlements and refugee camps and adequately allocating resources for the people in them. Nevertheless, remote sensing approaches to population mapping suffer from the above-mentioned issues of accuracy, mainly due to the need of VHR multispectral images for recognizing man-made structures and their extension, as well as single residential structures. But the precision obtainable by current systems is not comparable with existing GIS data for these areas, which are usually derived from much coarser (or more aggregated) sources. Alternatively, sparse or very sparse point-wise information may be available. This also offers a good example of the scale issues, since these different sources of information had to be considered on different levels of aggregation, and the fusion must necessarily be devised in a multiscale environment.

As this is the general framework of the problem, we should say that there is at the moment no global solution. There are instead a good number of papers in the technical literature addressing some of these issues and providing partial solutions. They are often based on *ad hoc* solutions, and do not consider more than one remote sensing and one GIS source at the same time, which makes them not really applicable to many situations. The most important and widely known methodology is that studied in Elvidge *et al.* (1996) and Sutton *et al.* (1997), based on the use of night-time satellite imagery from the DMSP satellites. A correlation between these measurements and population data has been found, and this may be improved using more remote sensing sources or land use information (Elvidge *et al.*, 1999), but it still needs to be carefully tuned for different parts of the world. As a matter of fact, one may expect that the correlation is different because of different ways of life, availability of electrical power and so on. The fusion with GIS data containing this information may be a possible solution, but this would be possible only using a multisource integration process. GIS data are made available by different administrations and have different legends, resolutions and other attributes. So, it is a really challenging problem still to be addressed.

One good example of a project addressing these issues is the Urban–Rural Database project, proposed and developed by the Centre for International Earth Science Information Network, Columbia University, in collaboration with the International Food Policy Research Institute and the World Bank. The aim of the project is to use night-time imagery from meteorological satellites to derive urban boundaries and connect them with population databases. In this project, nocturnal lighting is used to provide an estimate of the population by showing its activity in industrialized countries. In particular, this work (Pozzi *et al.*, 2003; Pozzi and Small, 2002) provides a grid of aggregated population data with a global coverage, using urban area boundaries detected from satellite imagery.

3.4 Present and future solutions

The problems introduced and discussed in the previous section have been addressed in different ways in the technical literature. At the moment there is no emerging methodology that looks like the ultimate approach. Each of the proposed solutions has its advantages and drawbacks, and we will try to provide a list of them after a brief introduction. Moreover, we must say that there is no place where all of the above-mentioned problems have been discussed in detail and addressed in a comprehensive framework. Therefore, the issues introduced in this chapter can be thought of as a series of windows overlooking a complex landscape, each window offering the reader a different view of the same problem and a possible solution to that particular view. The general framework for the fusion problem may be inferred from these partial views, and therefore will be provided at the end of this excursus, in the Conclusions section. Finally, some of the proposed solutions are just a glimpse of what is treated in more detail in other chapters of this book. They are considered here because they are mandatory for the characterization of the data fusion framework. The complete discussion of these issues is to be found in those chapters.

3.4.1 Multiscale analysis

The first characteristic of an effective methodology able to integrate GIS and remote sensing imagery is that it must be capable of working at different scales. The simplest translation of this concept is that the imagery should be decomposed, for instance by a pyramidal approach, into different copies at different ground resolutions. Working with these coarser copies of the original data may help in combining information coming from GIS layers, which may have different spatial accuracy, or make available aggregated data with a different spatial distribution in the scene.

This may be useful, for instance, in the extraction of objects that are of different sizes but maintain certain topological characteristics or relationships among themselves or with other objects. If finer resolution is used to analyse the scene, the interpretation must be connected through the scale (resolution) range, in addition to the spatial connections and (possibly) the temporal ones. Therefore, a merging strategy of the results at each resolution should be carefully designed. As an alternative, simultaneous extraction at higher resolution should be considered, which is, in any case, a more complex problem.

However, multiresolution is not the only way to cope with multiple scales in GIS–remote sensing fusion. A more refined approach involves taking into account the local scale of the objects in the scene, and adapting the analysis to this locally scaled neighbourhood. In this case, the combination of remotely sensed data and GIS may be adaptive to the context and be more efficient, since it depends directly on the interpretation of the image. This approach has been more recently proposed for image processing and remote sensing analysis in Gamba *et al.* (2004), where texture measures extracted from the co-occurrence matrix are computed using an adaptive window width, related to the local scale. The latter is computed using a semivariogram analysis in the local neighbourhood of each pixel, where the size is limited from the global scale of the whole scene. Note that in the cited paper, the approach was not used for GIS information combination, but it may be easily generalized for local scale characterization as a preliminary step to this combination. In particular, the procedure is based on three steps and is presented in Figure 3.1.

The two steps are, as mentioned before:

1. A *global-scale search*, aimed at defining the maximum spatial scale depicted in the image to be analysed.
2. A *local-scale search*, aimed at refining the scale analysis, looking only at the local neighbourhood of a pixel, in a search area defined by the previous step.

In the first step the whole image is considered as a unique environment, and the top level of the spatial scale range is set by a global scale analysis applied to the whole dataset. The reason for this step is the need to automatically obtain information about the dimensions of homogeneous texture patches in the image. The situation is complicated in remotely sensed data, because the scale of these features is naturally a function of the pixel size, and not only of their actual, real-world size. So, by looking at the whole remotely sensed image as a unique environment, we find the spatial scale of the land cover classes in the data. The second step allows a finer detailed analysis of the image in greater detail, but now we can limit ourselves to a smaller neighbourhood of each pixel, not larger than the global scale just extracted.

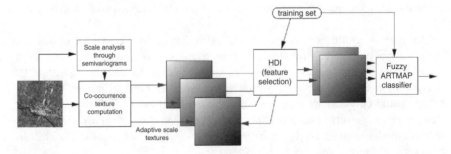

Figure 3.1 Multiscale automatic segmentation of a scene, using global as well as local scales, computed by means of the semivariogram analysis

For both scale analyses, the approach proposed in Townshend and Justice (1988), and used in Chen and Blong (2003) as a reference for the alternative wavelet algorithm, is implemented. In particular, the semivariograms are computed for the whole image or a subpart of it. The semivariograms are computed for any possible length, up to half the dimension of the studied area. Then, the maximum of the semivariogram curve is identified, and the corresponding length is the scale we are looking for.

3.4.2   Fusion techniques

The final piece of our puzzle is the definition of an algorithm where the above-mentioned characteristics may be integrated. Even for this step there is no absolute winner. Moreover, since this work is not aimed at providing a complete taxonomy of all the possible GIS–remote sensing fusion approaches, but just to give a general view of the most interesting algorithms, we will reduce our analysis to three main groups of techniques, as in the following list:

- *Techniques based on fuzzy logic*, aimed at taking care of the uncertainties and problems by a mathematical framework that models uncertainty through fuzziness, and based on which membership function is assigned to each data or information source.

- *Techniques based on non-parametric approaches*, such as neural networks, that do not assign any model to the data structures but instead adapt themselves to the different sources.

- *Techniques based on knowledge-based approaches*, which model the sources with respect to *a priori* knowledge about their importance, accuracy, and effectiveness for the proposed application.

Table 3.2 shows in compact form the main advantages and drawbacks of these three techniques. We should stress that inputs to the fusion algorithms are both remotely sensed imagery and GIS layers, but many of the following examples reduce the

Table 3.2   Taxonomy of fusion techniques

Technique	Advantages	Drawbacks
Fuzzy logic	Uncertainty model robustness	Careful tuning
Neural networks	Non-parametric Flexibility	Training required Complexity
Knowledge-based	Efficiency and accuracy Knowledge exploitation	Many rules *Ad hoc* solution

latter to raster format. This is done in order to compare the two information sets on the same spatial grid, usually defined by the spatial resolution of the remotely sensed data. Although this is a wise choice, it is also a reductive one. The raster and vector formats are suitable for different tasks and an algorithm needs to deal with both of them if it aims at their exploitation.

3.4.2.1 Fuzzy-based framework retaining accuracy information

Fuzzy logic is a powerful means to deal with different data sources, and may be adapted to combine GIS and remote sensing imagery or information. A first example may be found in Benz *et al.* (2004), where a fuzzy membership function, either chosen by default or user-defined, may be assigned to any information that is gathered about a segment of the original image. In this way the inherent uncertainties and problems in the different sources are mathematically introduced in a framework where each quality of an image object is used to compute joint membership of more complex or differently scaled elements of the scene. This approach is very interesting, but needs to be carefully tuned in order to model the fuzzy functions for each information source. Indeed, it allows tremendous flexibility, but also increases the time required to get meaningful and satisfying results.

A second example is the research proposed in Metternicht (2001). The authors evaluate the effects of changes in soil salinity, employing change detection techniques whose results are reclassified using fuzzy rules that discriminate among changes that are likely or unlikely to happen. The different degrees of likelihood of each couple are precisely modelled, using a particular set of rules based on fuzzy logic. The system is implemented into a GIS and uses for validation a number of information layers independently introduced in the same system. By using these rules, the procedure is able not only to define more precisely the change that has happened but also to quantify its likelihood and its extent, so that it may be used as input to further environmental models.

3.4.2.2 Non-parametric approaches

Other interesting approaches to data or information fusion are all non-parametric approaches. Among them, neural networks are by far the most commonly used. These networks do not need a model of the data source and may be tuned to different inputs by a suitable training phase.

Neural networks are used in Teng and Fairbairn (2002) to recognize objects that have been individuated in a high-resolution scene and assign them to one of the possible classes. To this end, their geometrical properties (area, perimeter, moments . . .) are considered and compared to those of already known objects in the same scene available in a GIS data layer. The advantage of neural networks in this case is their ability to cope with inputs having very different ranges and statistical properties. A very similar methodology is discussed in Rigol-Sanchez *et al.* (2003), where artificial neural networks are used to exploit GIS and remotely sensed data. The GIS data are translated into raster information with the same spatial resolution. The methodology proposed in the paper uses a feed-forward backpropagation network to individuate locations of interest for gold extraction, integrating information on land cover with knowledge of the slope and underground structure of the soil.

Finally, an interesting solution to problems caused by the different horizontal and positional accuracy of GIS and remotely sensed data is proposed in Mas (2004). Here the boundaries of a known area, coming from GIS layers, are *fuzzyfied* using a low pass filter. The application in this case is the monitoring of a tropical coastal area.

3.4.2.3 *Knowledge-based approaches*

Knowledge-based approaches, which are not inconsistent with fuzzy logic and are often coupled with them, are a valid alternative to non-parametric approaches. They require that parameters of the problem to solve or the analysis to perform be known *a priori*. This knowledge is modelled and fed into the classification or combination algorithm, with the precise aim of getting a particular result. In other words, while non-parametric approaches are a general tool that adapts to each situation, knowledge-based approaches are tailored to a given data structure and combination algorithm. The efficiency is therefore higher, but there is less flexibility.

A good example in this respect is Cohen and Shoshany (2002), where a crop-recognition system for the Mediterranean area was designed. The problem with this particular area was connected with the high level of segmentation of the rural area into different crops, so that high spatial variability was coupled with high spectral variability. A knowledge-based approach to crop recognition, based on multitemporal imagery (Landsat TM data translated into NDVI values) and rules for split and merge and generalization steps, successfully provided accurate maps. The approach exploited GIS layers, such as precipitation levels and soil types, for defining the rules by which to cluster or subdivide a large number of classes extracted using a fast unsupervised technique applied to an NDVI image sequence.

In a completely different application field, a similar approach was introduced for climate and air quality planning in urban and regional areas (Fehrenbach *et al.*, 2001). In this paper, land-use classes, aggregated into areal types, and ventilation situations were analysed to provide a classification of planning objectives. In other words, planning objectives for climate planning were translated into rules, and these rules were then reduced to a number of relationships between areal types and ventilation conditions, using two spatial grids, one of 25 m for urban areas and one of 100 m for regional analysis. The methodology was used to provide maps useful to planners whose objectives might be to 'improve ventilation', 'reduce the risks of strong winds', 'reduce air pollution' or 'improve the heat load situation'.

A final note on knowledge-based approaches comes from the results shown in Sader *et al.* (1995). In this paper a comparison among four different classification algorithms was provided, with the aim of providing a forest wetland inventory.

However, these algorithms were GIS- or remote sensing-based, and the only data fusion step was due to the fact that the information in the GIS layers, to be classified using a knowledge-based approach, was extracted from remotely sensed data. The comparison shows that there is no clear advantage in using knowledge-based approaches. There are always *optimized* and supervised statistical classification algorithms able to reach the same overall accuracy values. Additionally, this paper may be used to understand why GIS–remote sensing data fusion (using knowledge-based approaches or not) is useless if the information does not come from independent sources.

3.5   Conclusions

The scheme of the complete framework integrating the solutions in Section 3.4 is shown in Figure 3.2. As discussed above, it integrates the multiscale requirements of a good GIS–remotely sensed data fusion practice and the use of information, i.e. geometric or region primitives, for a more precise exploitation of the two sources.

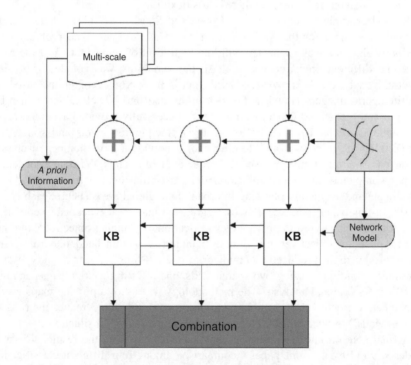

Figure 3.2 The general GIS–remote sensing fusion methodology proposed in this chapter

3.5.1 Integration of remote sensing and GIS into a change detection module

As an example of the system defined in a general way in the previous paragraphs, we offer here a very general change detection routine, integrating GIS layers of linear features into a change detection processing module. It should be noted that the procedure is general in the sense that the algorithm is applicable to change detection using remote sensing data only, and no requirements are made on the resolution of the imagery used. Moreover, the algorithm is written to integrate line detection and pixel-based change-detection techniques, and may be reduced to the one or the other part without losing efficiency. However, we stress that, following the discussion in section 3.2, it is easy to conclude that using both sources of information will be more effective.

The methodology, shown in Figure 3.3, therefore builds (possibly) on the results of a pixel- or even region-based change-detection algorithm, such as those reviewed in Coppin *et al.* (2004). It works towards improvements of the results obtained by using other features, and exemplifies this step by looking at linear features, such as those considered in the street extraction problem. In this procedure we propose to let the user choose whether he/she prefers to analyse the scene towards the lens of the linear features extracted, or to jointly consider the linear features and the classification or change maps. This module may be useful for change analysis starting from an already available GIS layer, feature extraction from remote sensing imagery, and change detection using classification and feature extraction routines in a joint effort for a more reliable output.

While pixel- or segment-based change detection methods are well defined, a change-detection algorithm made by linear and curvilinear feature comparison requires the definition of a methodology to compare two sets of these features. To this end, the comparison routine has been devised. It is devoted to determining the correspondences between linear features in two sets of line elements, either extracted from two images acquired at different dates, or already available at one of the dates in a GIS layer. Assuming that linear and curvilinear elements are in these sets, a very simple procedure for this comparison works as follows:

1. For each segment from the first group, starting from its extremes we build a box around it with transversal dimensions equal to δ pixels (the parameter δ is chosen by the user).

2. Then, the routine computes the number of pixels of any segment of the second group that fall into that box.

3. This allows us to compute association by using these values (if the above mentioned percentage is higher than a given threshold).

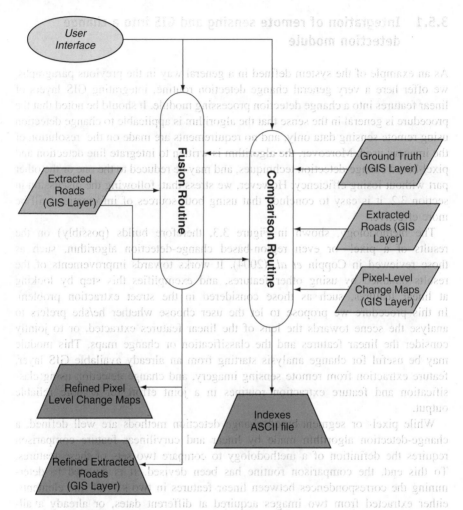

Figure 3.3 Block diagram of the feature-comparison and change-detection procedure

By this procedure it is possible to build a so-called *correspondence matrix*, in which we could describe the common segments present in several extractions. In this matrix, each extracted segment is associated with all the overlapping ones (following the above-mentioned criteria) in any other set.

If there is a GIS or manually extracted road layer available, a quantitative evaluation of the extracted dataset may be computed, considering two indexes (Wiedemann *et al.*, 1998; Wiedemann and Ebner, 2000), completeness and correctness. In order to improve change detection, it is possible to exploit both the information from the extracted features and those available as output of the change detection by the classification step. The procedure requires a fusion algorithm at the feature level

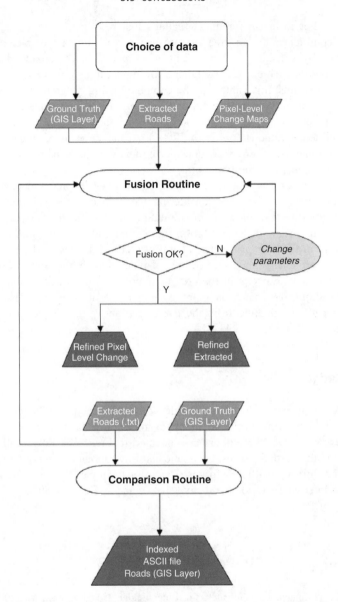

Figure 3.4 Flow diagram of the feature-comparison and change-detection procedure

between road segments extracted using imagery-dedicated routines and the land cover classification maps/ GIS layers of the same area.

Change detection may therefore be improved locally by comparison of more features, combined by the 'extraction confidence level' or the 'classification confidence'. For instance, for linear features that belong to road areas, the classification confidence for an area of the map surrounding an extracted element is computed by considering the percentage of pixels of the 'road' class in the neighbourhood of each extracted segment. Moreover, to take into account different possible road widths, this value may be computed using different window widths. The combination of the extraction and classification levels may therefore be used to test all the possible road widths and choose for each segment the one with the largest confidence value (Lisini *et al.*, 2004). At the same time, change detection classification may be improved by considering features extracted from the pre- and post-event (or generally, multitemporal) datasets.

After the comparison routine, the detected changes in pre- and post-segment datasets are recharacterized by considering the map of per-pixel changes, with the same algorithm that allows the exploitation of classification maps for feature extraction. This means that a change in the feature and a change in the maps are combined, in order to improve the reliability of the final result. This is done by using the feature change as a seed in the per-pixel change map and evaluating the percentage of change pixels in the area of interest. If this value is above a user-defined threshold, the change is approved; otherwise, it is rejected. The possible processing paths originating from the use of the above-mentioned procedure are depicted in detail in Figure 3.4.

Acknowledgements

This work would have not been possible without the help of Gianni Lisini and Giovanna Trianni, who provided many examples and helped in the design of the figures. Moreover, the authors are indebted to Francesca Pozzi and to many colleagues in the Earth Sciences, Geomatic and Environmental Engineering Departments of the University of Pavia for discussions on the topic. This work was partially funded by internal funds of the University of Pavia Department of Electronics, University of Pavia.

References

Benz, U. C., Hofmann, P., Willhauck, G., Lingenfelder, I. and Heynen M. (2004) Multi-resolution, object-oriented fuzzy analysis of remote sensing data for GIS-ready information. *ISPRS Journal of Photogrammetry and Remote Sensing* **58**, 239–258.

Bonnefon, R., Dherete, P. and Desachy, J. (2002) Geographic information system updating using remote sensing images. *Pattern Recognition Letters* **23**, 1073–1083.

Chen, K. and Blong, R. (2003) Identifying the characteristic scale of scene variation in fine spatial resolution imagery with wavelet transform-based sub-image statistics. *International Journal of Remote Sensing* **24**, 1983–1989.

Chica-Olmo, M., Abarca, F. and Rigol, J. P. (2002) Development of a decision support system based on remote sensing and GIS techniques for gold-rich area identification in SE Spain. *International Journal of Remote Sensing* **23**, 4801–4814.

Cohen, Y. and Shoshany, M. (2002) A national knowledge-based crop recognition in the Mediterranean environment. *International Journal of Applied Earth Observation and Geoinformation* **4**, 75–87.

Coppin, P., Jonckheere, I., Nackaerts, K., Muys, B. and Lambin, E. (2004) Digital change detection methods in ecosystem monitoring: a review. *International Journal of Remote Sensing* **25**, 1565–1596.

Elvidge, C. D., Baugh, K. E., Kihn, E. A. and Davis, E. R. (1996) Mapping city lights with nighttime data from the DMSP operational linescan system. *Photogrammetric Engineering and Remote Sensing* **63**, 727–734.

Elvidge, C. D., Baugh, K. E., Dietz, J. B., Bland, T., Sutton, P. C. and Kroehl, H. W. (1999) radiance calibration of DMSP-OLS low-light imaging data of human settlements. *Remote Sensing of Environment* **68**, 77–88.

European Commission (2004) *Establishing an Infrastructure for Spatial Information in the Community (INSPIRE)*. Communication COM (2004) 516, 23 July 2004.

Fehrenbach, U., Scherer, D. and Parlow, E. (2001) Automated classification of planning objectives for the consideration of climate and air quality in urban and regional planning for the example of the region of Basel/Switzerland. *Atmospheric Environment* **35**, 5605–5615.

Gamba, P. and Dell'Acqua, F. (2004) Monitoring urban areas for environment and security through remote sensing. *International Archives of Photogrammetry and Remote Sensing* **35**(B1), Istanbul, 12–23 July 2004, 319–324.

Gamba, P., Dell'Acqua, F. and Trianni, G. (2004) Automatic definition of scale features for remote sensed image interpretation. Proceedings of the Pattern Recognition in Remote Sensing 2004 Workshop, Kingston-upon-Thames, 27 August 2004 [unformatted CD-ROM].

Guindon, B. (1997) Computer-based aerial Image understanding: a review and assessment of its application to planimetric information extraction from very high resolution satellite images. *Canadian Journal of Remote Sensing* **23**, 38–47.

Hall, O., Hay, G., Bouchard, A., and Marceau, D. J. (2004) Detecting dominant landscape objects through multiple scales: an integration of object-specific methods and watershed segmentation. *Landscape Ecology* **19**, 59–76.

Hughes, G. F. (1968) On the mean accuracy of statistical pattern recognizers. *IEEE Transactions on Information Theory* **14**, 55–63.

INSPIRE. http://www.ec-gis.org/inspire/ [accessed 14 March 2005].

Janssen, L. L. F. and Molenaar, M. (1995) Terrain objects, their dynamics and their monitoring by the integration of GIS and remote sensing. *IEEE Transactions on Geoscience and Remote Sensing* **33**, 749–758.

Lisini, G., Tison, C., Cherifi, D., Tupin, F., and Gamba, P. (2004) Improving road network extraction in high resolution SAR images by data fusion. Proceedings of CEOS SAR Workshop 2004, Ulm, Germany, May 2004 [unformatted CD-ROM].

Maksud Kamal, A. S. M. and Midorikawa, S. (2004) GIS-based geomorphological mapping using remote sensing data and supplementary geoinformation: a case study of the Dhaka city area, Bangladesh. *International Journal of Applied Earth Observation and Geoinformation* **6**, 111–125.

Mas, J. F. (2004) Mapping land use/cover in a tropical coastal area using satellite sensor data, GIS and artificial neural networks. *Estuarine, Coastal and Shelf Science* **59**, 219–230.

Mason, S. O., Baltsavias, E. P. and Bishop, I. (1997) Spatial decision support systems for the management of informal settlements. *Computer, Environment and Remote Sensing* **21**, 189–208.

Meeks, W. L. and Dasguota, S. (2004) Geospatial information utility: an estimation of the relevance of geospatial information to users. *Decision support Systems* **38**: 47–63.

Metternicht, G. (2001) Assessing temporal and spatial changes of salinity using fuzzy logic, remote sensing and GIS. Foundations of an expert system. *Ecological Modelling* **144**, 163–179.

Peddel, D. R. and Ferguson, D. T. (2002) Optimization of multisource data analysis: an example using evidential reasoning for GIS data classification. *Computers and Geosciences* **28**, 42–52.

Pozzi, F. and Small, C. (2002) Vegetation and population density in urban and suburban areas in the USA. Proceedings of the Third International Symposium of Remote Sensing of Urban Areas, 11–13 June 2002, Istanbul, Turkey, 489–496.

Pozzi, F., Small, C. and Yetman, G. (2003) Modelling the distribution of human population with night-time satellite imagery and gridded population of the world. *Earth Observation Magazine*.

Prol-Ledesma, R. M., Uribe-Alcantara, M. and Diaz-Molina, O. (2002) Use of cartographic data and Landsat TM images to determine land use change in the vicinity of Mexico City. *International Journal of Remote Sensing* **23**, 1927–1933.

Rao, M., Pandey, A., Ahuja, A. K., Ramamurthy, V. S. and Kasturirangan, K. (2002) National Spatial Data Infrastructure – coming together of GIS and EO in India. *Acta Astronautica* **51**, 527–535.

Rigol-Sanchez, J. P., Chica-Olmo, M. and Abarca-Hernandez, F. (2003) Artificial neural networks as a tool for mineral potential mapping with GIS. *International Journal of Remote Sensing* **24**, 1151–1156.

Sader, S. A., Ahl, D. and Liou, W. S. (1995) Accuracy of Landsat TM and GIS rule-based methods for forest wetland classification in Maine. *Remote Sensing of Environment* **53**, 133–144.

Shahshahani, B. M., and Landgrebe, D. A. (1994) The effect of unlabeled samples in reducing the small sample size problem and mitigating the Hughes phenomenon. *IEEE Transactions on Geoscience and Remote Sensing* **32**, 1087–1095.

Smith, G. M. and Fuller, R. M. (2001) An integrated approach to land cover classification: an example in the Island of Jersey. *International Journal of Remote Sensing* **22**, 3123–3142.

Stefanov, W. L., Ramsey, M. S. and Christensen, P. R. (2001) Monitoring urban land cover change: an expert system approach to land cover classification of semiarid to arid urban centres. *Remote Sensing of Environment* **77**, 173–185.

Sutton, P., Roberts, D., Elvidge, C. and Meij, H. (1997) A comparison of nighttime satellite imagery and population density for the continental United States. *Photogrammetric Engineering and Remote Sensing* **63**, 1303–1312.

Tapiador, F. J. and Casanova, J. L. (2002) Land use mapping methodology using remote sensing for the regional planning directives in Segovia, Spain. *Landscape and Urban Planning* **62**, 103–115.

Teng, C. H. and Fairbairn, D. (2002) Comparing expert systems and neural fuzzy systems for object recognition in map dataset revision. *International Journal of Remote Sensing* **23**, 555–567.

Townshend, J. R. G. and Justice, C. O. (1988) Selecting the spatial resolution of satellite sensors required for global monitoring of land transforms. *International Journal of Remote Sensing* **9**, 187–236.

Varma, H., Fadaie, K., Habbane, M., Stockhausen, J. (2003) Confusion in data fusion. *International Journal of Remote Sensing* **24**, 627–636.

de la Ville, N., Chumaceiro Diaz, A. and Ramirez, D. (2002) Remote sensing and GIS technologies as tools to support sustainable management of areas devastated by landslides. *Environment, Development and Sustainability* **4**, 221–229.

Wald, L. (1998) A European proposal for terms of reference in data fusion. *International Archives of Photogrammetry and Remote Sensing* **32**, 651–654.

Walter, V. (2004) Object-based classification of remote sensing data for change detection. *ISPRS Journal of Photogrammetry and Remote Sensing* **58**, 225–238.

Wiedemann, C., Heipke, C., Mayer, H. and Jamet, O. (1998) Empirical evaluation of automatically extracted road axes. In Kevin J., Bowyer, P. and Jonathon P. (eds), *Empirical Evaluation Methods in Computer Vision*. IEEE Computer Society Press: Los Alamitos, CA, USA; 172–187.

Wiedemann, C. and Ebner, H. (2000) Automatic completion and evaluation of road networks. *International Archives of Photogrammetry and Remote Sensing* **33**(3B): 979–986.

Xiuwan, C. (2002) Using remote sensing and GIS to analyze land cover change and its impacts on regional sustainable development. *International Journal of Remote Sensing* **23**, 107–124.

Xue, Y., Cracknell, A. P. and Guo, H. D. (2002) Telegeoprocessing: the integration of remote sensing, geographic information system (GIS), global positioning system (GPS), and telecommunications. *International Journal of Remote Sensing* **23**, 1851–1893.

4

The importance of scale in remote sensing and GIS and its implications for data integration

Peter M. Atkinson

School of Geography, University of Southampton, UK

4.1 Introduction

An understanding of the effects of sampling upon acquired data is a prerequisite for intelligent integration of remotely sensed imagery with other spatial data within a geographical information system. The sampling framework, defined as the set of all parameters that determine how data are acquired on a property of interest, has important consequences for the nature of the values and, thus, the spatial variation in the resultant variable. In particular, the sampling framework determines in part the scales of spatial variation that are realized in the variable that is presented for analysis (Moellering and Tobler, 1972; Atkinson and Tate, 2000). This is as true for GIS vector data as it is for remotely sensed imagery and other raster datasets. It follows that any attempts to integrate remotely sensed imagery with spatial data within a GIS should include careful consideration of: (a) the sampling frameworks used to acquire spatial data; (b) the scales of variation present in spatial data; and (c) the consequences of these for combining the original spatial variables.

The objective of this chapter is to review the various component effects summarized above to arm the researcher or GIS user with a conceptual framework with

Integration of GIS and Remote Sensing Edited by Victor Mesev
© 2007 John Wiley & Sons, Ltd.

which to understand the consequences of attempting data integration and an aware-ness of the possibilities for improving the methods used for integration. The focus of this chapter is on handling remotely sensed images, in combination with GIS vector and raster data, reflecting the interests of the author. The books edited by Quattrochi and Goodchild (1997) and Tate and Atkinson (2001) provide a range of examples of the importance of scale in remote sensing and GIS, in a broader context.

Section 4.2 provides an introduction to the concept of scale of measurement. Section 4.3 introduces the concept of scale of spatial variation. Section 4.4 then considers some of the issues introduced in Sections 4.2 and 4.3 within the context of GIS data integration. Section 4.5 provides a brief conclusion.

4.2   Data models and scales of measurement

It is important when considering issues of scale to distinguish between the scales of measurement and the scales of spatial variation that are observed in data. This section deals with scales of measurement.

4.2.1   Raster imagery

4.2.1.1   Raster imagery and the RF model

A remotely sensed image is an example of data acquired in the raster data model (i.e. an image, see Figure 4.1). The raster data model is comprised of several non-overlapping areal units called pixels, arranged on a regular grid (Burrough and McDonnell, 1998). The sampling framework of an image is relatively easy to conceptualize. One can imagine a mesh being placed over a continuously varying space and all variation within each pixel being averaged out to produce a set of values, one per pixel. In this sampling process, much variation is lost. This sampling process is described in more detail below. The important point to make here is that the sampling framework of a remotely sensed image is fixed prior to data acquisition. For this reason, the pixels bear very little or no relation to the properties of the scene. Note that in some circumstances it may be possible to vary the pixel size locally (Csillag *et al.*, 1996).

The raster data model can be used to describe continuous variation (in either continuous or categorical variables) or objects. In the former case, the space that the image covers is commonly represented as a random function (RF). In an RF each observation is modelled as a draw from a cumulative distribution function (CDF) and each CDF is conditioned on others (e.g. neighbours) to account for dependence between observations. Commonly, a variogram or spatial co-variance function is chosen as a suitable parametric model to define the spatial dependence between neighbours (Journel and Huijbregts, 1978). An example variogram is shown in

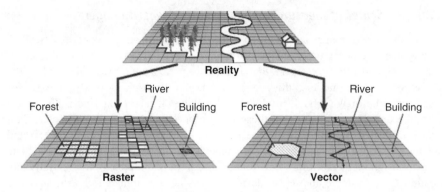

Figure 4.1 The raster and vector data models

Figure 4.2. It plots semivariance (a measure of dissimilarity) against lag **h** (the vector distance and direction between a pair of observations, i.e. the separation). The form of variogram shown in Figure 4.2 is typical of variation encountered in remotely sensed images: as **h** increases, the dissimilarity increases either asymptotically towards, or reaching, a maximum.

In the latter case, an object-based view of the world replaces the RF: pixels are conjoined into larger agglomerations of pixels to represent objects that are perceived to exist in reality (e.g. house, road, forest stand). The raster data model is most naturally combined with the RF model, primarily because the pixel locations bear no relation to (i.e. are not conditional upon) the actual scene objects.

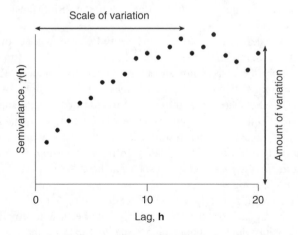

Figure 4.2 An example of a typical experimental variogram

4.2.1.2 Scales of measurement in remotely sensed imagery

Remotely sensed images are acquired through a sampling framework, as described briefly above. Conceptually, it is useful to distinguish between the scene of interest, the sensor and the atmosphere that lies between them. Of particular use is the definition of four models: the scene model, the atmosphere model, the sensor model and the image model (Strahler *et al.*, 1986; Curran *et al.*, 1998). The image that is acquired is a function of these three other components. In particular, light that is reflected or emitted from the Earth's surface (i.e. scene) may be affected by the atmosphere (scattered and attenuated) before it reaches the sensor. The sensor design then determines how that altered light is sampled, both spectrally and spatially. For example, the sensor design determines the number and position of the wavebands that are recorded from within the electromagnetic spectrum. A common set of wavebands (e.g. for fine spatial resolution sensors such as IKONOS and Quickbird) is the blue (0.45–0.52 μm), green (0.52–0.6 μm), red (0.63–0.69 μm) and near-infrared (0.76–0.9 μm) wavebands (Aplin *et al.*, 1997). Of more interest, given the present context, is the spatial sampling afforded by the sensor design. Sensors operate physically in several ways (e.g. push-broom, side scanning) but the consequence of that operation in general terms is the same – an image.

In remote sensing, the term 'spatial resolution' is used interchangeably with 'pixel size'. However, there are some differences and to set them out we need to introduce a new term borrowed from geostatistics – the support (Journel and Huijbregts, 1978). The support is the size, geometry and orientation of the space over which an observation is defined. In remote sensing, the support size is determined by the instantaneous field of view (IFOV) and the flying height of the sensor platform. The shape of the support is not square, but is generally centre-weighted due to the point-spread function (PSF) of the sensor. The exact shape and the extent of centre-weighting are determined by the sensor specification. Importantly, the observations in a remotely sensed image generally overlap due to the PSF (Manslow and Nixon, 2002) (Figure 4.3).

Given the above definition of a remotely sensed image as a set of overlapping observations, it is possible to define the spatial resolution and pixel more precisely. The pixel is simply the cell within an output device to which an observation or value within an image is assigned. So, whereas the observations overlap, pixels abut (Figure 4.3). The spatial resolution is a function of both the support and the sampling density. Thus, if observations were obtained more sparsely, such that they did not cover the space of interest, the spatial resolution would decrease, even though the support would remain the same. In remote sensing, the pixel size and spatial resolution are often approximately the same because the pixel size determines the sampling grid spacing.

The above view of a remotely sensed image as a sampled version of the atmospherically altered signal from the scene is important because it provides a conceptual framework with which to understand scale. All data on the environment are

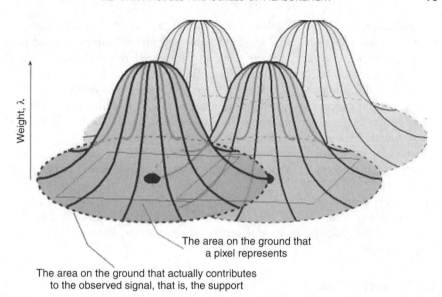

Weight, λ

The area on the ground that
a pixel represents

The area on the ground that actually contributes
to the observed signal, that is, the support

Figure 4.3 Illustration of the centre-weighted supports of remotely sensed image observations overlapping spatially

acquired through some sampling framework, such that the data are a function of both the property of interest (in this case the scene) and the sampling framework (as determined by the sensor characteristics) (Figures 4.3, 4.4). For a remotely sensed image, the size of support, approximated by the pixel size, is the most important parameter of the sampling framework (Atkinson and Tate, 2000). The support can be thought of as representing a primary scale of measurement, which may be conceived of as a filter on reality (Figure 4.3).

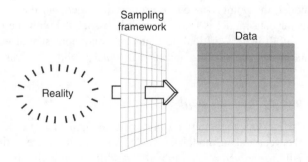

Figure 4.4 Spatial data seen as a function of some underlying reality and the spatial sampling framework

Importantly, the scales of spatial variation that are captured in the data are a function of the scales of spatial variation in the property of interest defined on a point support (i.e. in reality) and the support (note that relative to the above model we now neglect the effect of the atmosphere in order to simplify the exposition). This statement presupposes that we could measure on a point support, which we cannot, but the concept is important: an external reality is defined such that the scales of spatial variation detectable in observed data are a function of those in reality filtered by a given sampling framework.

If the support is varied, then the scales of spatial variation detectable in the data will change. If the support is reduced in size, then more (fine-scale) variation will be revealed. If the support is increased, then some variation will be lost (leaving a greater proportion of coarse-scale variation). The important point is that if data are to be combined within a GIS, then we should have a good idea (i.e. a model) of the effect of the sampling framework and the support on the datasets that are to be integrated together.

4.2.2   Vector data

4.2.2.1   Vector data and the object-based model

The vector data model is fundamentally different from the raster data model in several ways. First, the observations in the vector data model are comprised of points, i.e. $\mathbf{x} = (x, y)$ location in a Cartesian coordinate system. These points are usually placed so as to represent objects in the scene of interest. Point objects are represented with a single point, lines with multiple points and area objects with a sequence of multiple points in which the first and last point are the same (Figure 4.1). Generally, the points are placed so as to define the border or geometry of the objects. The objects, once represented, can be attributed with values of a particular variable (e.g. land use).

Clearly, a fundamental difference between the raster and vector data models is that in the raster data model, the pixels bear no relation to the scene, whereas in the vector data model, the point locations *represent* the objects in the scene. For this reason, the vector data model is most commonly associated with the object-based view of the world. It is very difficult to represent continuous spatial variation with the vector data model because the objects are depicted with zero variation *within* them.

4.2.2.2   Scales of measurement

It may seem at first that the sampling principles as applied to raster imagery do not apply to the vector data model; however, the principles are entirely general and apply equally to the vector data model. The difference is that the space being sampled is no longer a continuous Euclidean space, but the space defined by the boundary (geometry) of the object. Sampling more densely (e.g. digitizing more frequently when representing, for example, a railway line) reveals greater variation in the curvature of that feature. Sampling less densely provides a more generalized representation. The issue of line and feature generalization (i.e. how to coarsen the scale of representation in such a way as to produce a cartographically consistent map) has been the subject of much research in GIS (e.g. Ware and Jones, 1998).

A rather more complex problem is the so-called modifiable areal unit problem (MAUP) associated with census and similar datasets (Openshaw, 1984; Amrhein and Wong, 1996). For census data, not only does the scale of measurement vary (the aggregation problem), but also the actual realization of a particular sampling model (with fixed parameters such as size of support) can have large effects on the observed data. This effect is referred to as the zonation problem (Martin, 1996). In addition, and with serious consequences for the ability to apply statistical procedures, the actual units vary from place to place in their size, geometry and orientation (Atkinson and Martin, 1999).

4.3 Scales of spatial variation

Having considered scales of measurement in the previous section, the goal in this section is to describe quantitatively the scales of spatial variation present in data. The focus in this section is on the raster data model and random function view of the world.

4.3.1 Spatial variation in raster data

It should be apparent from the preceding sections that the scales of spatial variation that are detectable in data are a function of the sampling framework and those present in reality. The question is, how should one characterize the scales of spatial variation present in data? To answer this question, it is necessary to define what is meant by 'scales of spatial variation'.

4.3.1.1 Characterizing scales of spatial variation

Using the object-based model one can imagine several discs placed on a background (such discs might represent tree crowns in a remotely sensed scene). Large discs would represent a large-scale of spatial variation relative to small discs. Adopting the continuous field model of spatial variation, imagine two scenarios involving variation in the biomass of vegetation: (a) a tropical rainforest in which the spatial variation in biomass is clumped producing a rough texture; and (b) a field of pasture in which the smoothly varying underlying soil patterns result in a smooth texture in biomass. In (a) the variation is fine-scale or high-frequency, and in (b) the variation is coarse-scale or low-frequency. These loose definitions help to conceptualize the meaning of 'scales of spatial variation', but do not provide a quantitative means to measure them.

The variogram, the central tool of geostatistics, is one of several functions that may be used to characterize the scales of spatial variation present in spatial data (Goovaerts, 1997; Atkinson, 1999; Chilès and Delfiner, 1999). Other approaches such as wavelets are also common (Mallat, 1989; Chen and Blong, 2003). The variogram is particularly suited to raster imagery and the RF model as described above, but equivalent functions can be defined for vector data.

The experimental or sample variogram is a plot of several empirical values of semivariance against several discrete lags (Figure 4.2). The experimental vari-ogram, $\hat{\gamma}(\mathbf{h})$, can be estimated from $p(\mathbf{h})$ paired observations, $z(\mathbf{x}_\alpha)$, $z(\mathbf{x}_\alpha + \mathbf{h})$, $\alpha = 1, 2, \ldots p(\mathbf{h})$ using:

$$\hat{\gamma}(\mathbf{h}) = \frac{1}{2p(\mathbf{h})} \sum_{\alpha=1}^{p(\mathbf{h})} \{z(\mathbf{x}_\alpha) - z(\mathbf{x}_\alpha + \mathbf{h})\}^2 \qquad (4.1)$$

To use the variogram in most geostatistical procedures, it is necessary to fit a mathematical model to the empirical values. There are certain rules governing the choice of model (models must be 'authorized') and its fitting that need not concern us here (see McBratney and Webster, 1986). Gstat is an easy-to-use soft-ware package that allows variogram estimation and model fitting (Pebesma and Wesseling, 1998).

Each permissible model is defined by several parameters. Most models are tran-sitive; they reach a maximum value of semivariance (the sill, c) at a specific lag (the range, a). Some models are unbounded; the semivariance increases indefinitely with lag. For transitive models the range defines the limit to spatial autocorrela-tion; at lags less than the range values they are correlated, but beyond it they are expected to be independent. In this sense, the range defines a maximum scale of spatial variation in the data. The form of the fitted variogram model defines the entire range of scales of spatial variation in the data.

Some of the most commonly used authorized variogram models are the expo-nential and spherical models. The exponential model is given by:

$$\gamma(h) = c \cdot \left[1 - \exp\left(-\frac{h}{d}\right)\right]. \qquad (4.2)$$

where d is the non-linear distance parameter. The exponential model reaches the sill asymptotically.

The spherical model is given by:

$$\gamma(h) = \begin{cases} c \cdot \left[1.5\frac{h}{a} - 0.5\left(\frac{h}{a}\right)^3\right] & \text{if } h \leq a \\ c & \text{if } h > a \end{cases} \qquad (4.3)$$

4.3.1.2 Characterizing error

Most models fitted to experimental variograms include a further model called the nugget effect model, which is given by:

$$\gamma(h) = \begin{cases} 0 & \text{for } h = 0 \\ c_0 & \text{for } h > 0 \end{cases} \tag{4.4}$$

where c_0 is the nugget variance or sill of the nugget effect model. The nugget effect is essentially a completely flat model that takes the same value of semivariance, independent of lag, with the exception of $|\mathbf{h}| = 0$, which must always take the value of zero. Adding the nugget model means essentially that the semivariance is increased at all lags and that the intercept of the model with the ordinate is positive and equal to c_0.

The nugget variance is due to (a) micro-scale variation that has not been accounted for by the sampling and (b) measurement error. For remotely sensed imagery, the nugget variance is sometimes used to represent measurement error (Curran and Dungan, 1989; Atkinson *et al.*, 1996).

4.3.1.3 Upscaling and downscaling

'Upscaling' refers to an increase in the size of the support, whereas 'downscaling' refers to a decrease in the size of the support (Figure 4.5). It is possible to upscale a remotely sensed image readily through weighted averaging. One possible method is to simply average the existing pixels to create larger ones. This simple operation may be adequate for some purposes, but it is not a precise representation of a real remotely sensed image at the larger pixel size because it assumes a square wave response, which (from above) we know to be inappropriate, due to the effects of the PSF (e.g. Figure 4.3). Downscaling is more difficult but not impossible, as described in the example on super-resolution mapping later in this chapter.

Adopting the RF model in a geostatistical setting, it is possible to define a model of the effect of the support on spatial variation detectable in a remotely sensed image. More specifically, the geostatistical operation of regularization models

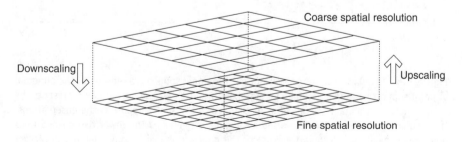

Figure 4.5 Upscaling and downscaling

the effect of the support on the variogram that characterizes the spatial variation (including the scales of spatial variation) in an image. The model is defined by Clark (1977) and Journel and Huijbregts (1978) as:

$$\gamma_v(\mathbf{h}) = \bar{\gamma}(v, v_{\mathbf{h}}) - \bar{\gamma}(v, v) \tag{4.5}$$

where $\bar{\gamma}(v, v_{\mathbf{h}})$ represents the integral punctual semivariance between two pixels of size v whose centroids are separated by $\mathbf{h}$, and $\bar{\gamma}(v, v)$ represents the integral punctual semivariance within a pixel of size v (i.e. the within-block variance).

The spatial variation observed in data (i.e. the variation *between* data) is equal to the spatial variation in the unobserved underlying property of interest minus the spatial variation that is averaged out or lost within the support. Figure 4.6 shows an example in which a punctual or point variogram model has been estimated by comparing the regularized or convolved semivariance values obtained from the punctual model, using equation 4.5, with the experimental semivariance values obtained on the observation support. The figure shows that the same punctual model, once estimated, can be used to estimate the variogram for any support larger than that of the original data (Jupp *et al.*, 1998, 1999). This model is important because it embodies some of the fundamental sampling concepts introduced in earlier sections (see Figure 4.4).

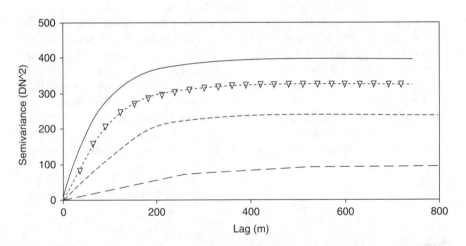

Figure 4.6 Regularization of the variogram. The upper solid curve represents the punctual or point support variogram (which may not exist in practice). The dotted line represents the same punctual variogram convolved or regularized to the same support as the observations, thus matching the experimental variogram (triangle symbols). The lower two dashed lines represent convolutions (regularizations) of the punctual model to supports that are larger than the original support. DN, digital number

4.3.2   Scales of variation in vector data

The principles outlined above in relation to the RF model also apply to the vector data model and the object-based view of the world. In the attribute sense, vector data can be seen as providing a realization of a categorical RF model for every point in a continuous space. Then the above principles apply directly. For example, it would be possible to discretize the vector data layer and estimate the experimental variogram, based on the discretized grid.

In the geometric sense of defining the boundaries of objects, the above principles apply less directly. It is true that the greater the sampling density along a line, the less generalized the line represented by the series of point data will be (given that sufficient detail exists in the original property being measured – for cartographic map data which are already an abstraction of reality, this may not be the case). However, the variogram is unlikely to be used to characterize spatial variation along a line. More commonly, object-based statistics, simple examples of which are compactness, roundness and convexity, have been applied to determine the degree of curvature in boundaries. The fractal model has been particularly useful in describing the scales of spatial variation (or, indeed, scale invariance, where this holds true) present in lines and areas whose dimension lies between some integer value.

Attempts to characterize the scales of spatial variation in census data (with the associated MAUP) have been limited (Atkinson and Martin, 1999; Fotheringham *et al.*, 2000). This is largely because the sampling imposed by the variable units (supports) limits the application of spatial statistics. Simple statistics, such as Moran's I and Geary's c, have been applied to census data to measure autocorrelation.

4.3.3   Processes in the environment

4.3.3.1   Processes and forms

An important distinction to make in our conceptual model is between form and process. Observations of the real world that can be stored as data ready for analysis (e.g. within a GIS) relate information on form (i.e. spatial variation or pattern). However, spatial forms in the real world are not static; they are constantly changing. The rules by which such changes occur are referred to as 'processes'. In the general sense, it is not possible to measure processes and so they must be inferred by analysing changes that occur in spatial forms.

Processes are the primary object of interest in science; knowledge of process encapsulates understanding and the consequent ability to explain. Therefore, although processes are unlikely to be the primary object of interest in the context of remote sensing and GIS data integration, they deserve attention.

4.3.3.2 Process modelling

An important and emerging field of investigation is spatially distributed dynamic modelling (SDDM), also referred to as GeoDynamics (Atkinson *et al.*, 2005). In such spatially distributed models, an area of interest is commonly represented as a grid or mesh, and the data within the cells so defined are allowed to vary through time as a function of some rules that code the processes of interest. Such models have been applied extensively within fields such as riverine flooding, geomorphology, landscape evolution, landscape ecology and infectious disease transmission amongst others (for examples, see Atkinson *et al.*, 2005). The objective of such modelling varies, but can include: (a) increasing understanding of the system; (b) forecasting, often in real-time using data assimilation techniques; and (c) evaluation of what-if scenarios. The latter objective includes evaluation of the behaviour and response of the system under future possible scenarios (e.g. climate change) and alternative management and planning scenarios. This ability to evaluate the system independent of new measurement should make clear why knowledge of process is so important and should not be overlooked, even in a book focused on form.

4.3.3.3 Scales of representation

Processes are represented in a model (e.g. SDDM) as a series of usually rather crude rules, which can be written either as mathematical expressions or in computer code. These rules are abstractions of the real processes (which often lie in the realm of physics). Further, the abstraction often occurs at a particular set of spatial and temporal scales, as determined by the spatial and temporal sampling used to acquire the data on which the rules have been calibrated and validated. A fair understanding of this abstraction is important because it determines the level to which objective (a) above is feasible – often it is not. Importantly, the processes underlying the measurable forms of interest will be defined at particular spatial and temporal scales. Whether these scales are represented adequately in a process model as described above is an important question.

4.4 Remote sensing and GIS data integration

In this section, several examples are given of the integration of datasets within a GIS. Now that the concepts of scales of measurement and scales of spatial variation have been introduced, the importance of such scales for data integration can be highlighted. Two operations are used as vehicles to drive the discussion: GIS overlay (and regression) and remote sensing classification.

4.4.1 Overlay and regression

Common objectives in GIS are site suitability and habitat suitability mapping. These often centre on the overlay operation applied to several input data layers, which may be provided in a variety of formats, including both the raster and vector data models. For example, selecting a suitable habitat for a rare species of bird may require image data on land cover (e.g. woodland density), but also vector data representations of features such as hedgerows, rivers and motorways.

Statistical regression, like overlay, requires data integration and within a GIS may require combination of data from quite disparate sources. In regression, the objective is to predict one (target) variable based on other correlated explanatory variable(s).

It should be clear from other chapters of this book that overlay is a rather crude procedure in GIS. It assumes expert knowledge that may not exist, or may be poorly founded, and does not involve any statistical fitting to empirical data. As a result, the output from an overlay analysis may be arbitrary and difficult to defend. Where data on the desired variable exist (e.g. data on suitable habitats), it is often preferable to fit a statistical model (e.g. multiple regression) and use the coefficients of the fitted model to predict suitability.

4.4.1.1   Scales of measurement

Whether data are to be combined using overlay or regression, the same concerns over the scales of measurement arise. Consider, first, the simple case where two raster data layers are to be combined. An example would be the integration of a remotely sensed image with a digital elevation model (Janssen et al., 1990; Janssen and Molenaar, 1995). These data layers may have different scales of measurement (e.g. spatial resolutions). A question then arises over whether it is necessary to change the spatial resolution of one dataset in order to match that of the other. A common choice would be to degrade the finer spatial resolution to match the coarser one. Whether or not this is a sensible strategy really depends on the scale(s) of spatial variation in the underlying property, and hence the resultant variable. Specifically, the investigator should ask whether the resultant spatial resolution is adequate to resolve the spatial variation of interest.

In another circumstance, the two raster datasets to be combined may be provided with the same spatial resolution. However, it does not follow automatically that the correlation between the two datasets will be a maximum when the support on which they are represented is the same. Consider the situation in which the explanatory variable is comprised of a source of variation of interest plus some other source of variation, which may be considered as spatially correlated noise that is uncorrelated with the target variable. If the scales (frequencies) of the two sources of spatial variation are different, then it will be possible to filter the explanatory variable to actually increase the correlation with the target variable (Goovaerts, 1997). This amounts to changing the scale of measurement such as to focus on the required variation.

4.4.1.2   Transformation

Often it is necessary to transform data into a compatible coordinate system and data model to facilitate comparison with other data layers. One of the key problems in GIS is that when data are transformed, the transformation procedure imparts a scale of measurement on the resulting data that is an artefact of the algorithm and not the original sampling incurred.

A common operation employed in transforming point data to the raster data model is interpolation. Many algorithms exist for spatial interpolation, including the widely applied inverse distance weighting (IDW) interpolation and the geostatistical technique of kriging. All common algorithms are smoothing operators, because the predicted value is simply a weighted average of neighbouring data. Thus, some variance is lost as a function of the interpolation procedure and the resulting predicted variable is smoother than the original. This effect can be particularly severe for algorithms such as IDW. Investigators are forced to apply such transformations because no alternative exists; however, they are often unaware of the consequences of such action.

In the above example, the smoothing imparted by the interpolation procedure amounts to an increase in the support of the predicted variable above that of the original variable. In fact, the support of the new variable is precisely equal to the original support convolved with the support defined by the weighted set of neighbouring points used in the interpolation, which approximates a sample of a distance-decay function of some description. This distance-decay function spreads out the support of the prediction beyond the support of the original data (Atkinson and Kelly, 1997; Atkinson and Tate, 2000). Users of interpolation procedures should be aware of the scale-related effects of the algorithms that they are applying.

In GIS operations such as regression, smoothing can lead to undesirable effects. For example, suppose that only the explanatory x variable has been smoothed and the predicted y variable has not been smoothed. Then the scatterplot between the two variables will be altered such that a best-fit line through the scatter will have an increased slope. Such a relation would not be applicable to other datasets that had not been smoothed. This is a key problem in remote sensing–GIS data integration.

The combination of raster and vector data through overlay or regression often involves the process of converting one form of data model to another. A common choice is to convert vector data to the raster data model (rasterization). This process of converting from one data model to another may involve adding a scale of measurement, as discussed above.

4.4.1.3   Geometric error

It is well known that geometric error has a large effect on procedures in a GIS that use multiple data layers. Consider the situation in which two vector data layers are to be combined. It is well known that geometric imprecision will lead to sliver polygons in the resultant vector output map.

It is interesting in the present context to consider the effect that generalization in the observed vector data has on the combined output. Suppose that one dataset is more generalized than the other. Then, one would expect sliver polygons to result from the imposed scale of measurement in the one dataset. The effect is similar to that of geometric error, but with a different spatial character. If both vector data layers have been obtained with the same levels of generalization, then the problem may diminish, but it may not vanish. There is still likely to be a component of error due to the spatial resolution or sampling density with which the lines have been represented.

The problems and issues relating to geometric error are even more important when the objective is change detection and monitoring (Westmoreland and Stow, 1992). Apparent 'changes' appear at the boundaries of objects simply because the objects have not been located precisely in either or all datasets. Such slivers at the boundaries of objects can have a severe effect on statistics such as the mean change vector ('vector' here meaning the change from one point to another in feature space).

A key question is, how accurate does the registration need to be? Well, let us quantify the geometric error using the standard deviation s, within which 68% of deviations are expected to lie. The *required* standard deviation will depend, in large part, on the scale of spatial variation in the data layers being used within the procedure. Take the example of overlay. If the objects of interest (assuming a scene model comprised of objects) were, say 300 m on a side (e.g. agricultural fields), then the s that may be tolerated would be larger than if the objects were around 10 m on a side (e.g. residential houses). These basic principles are similar to those that have been articulated for the scale of measurement (a smaller pixel being required to resolve smaller objects) (Woodcock and Strahler, 1987; Townshend and Justice, 1988; Atkinson and Curran, 1997).

The object-based view is useful to convey principles, but these principles apply also to continua. If the scale of spatial variation is fine (as characterized by the variogram range and model type), then a smaller s will be required. For example, if the investigator is interested in variation in, say, leaf area index (LAI) between tree crowns in a forest stand with a predominant scale of spatial variation (e.g. variogram range) caused by tree crowns, then the pixel size *and* geometric registration error will need to be much smaller than the variogram range.

It is interesting to note that the ability to perform a geometric rectification is often dependent on sufficient spatial variation in the image or dataset to be registered. Thus, registration precision is likely to be greater where the features of interest are sufficiently well resolved. It is for this reason that standard deviations of around one pixel or less are commonly reported for remotely sensed images, independent of the spatial resolution.

4.4.2 Remote sensing classification of land cover

Classification of land cover using remotely sensed imagery is commonly pixel-based (Tso and Mather, 2001). This means that each pixel is classified based on some algorithm, independently of its neighbours. Given that neighbouring pixels in a remotely sensed image are rarely independent, this does not seem a sensible strategy (Cushnie, 1987). There are many alternative algorithms that in one way or another make use of the correlation between neighbouring pixels to increase the precision with which land cover can be classified. A good example is the class of algorithms referred to as Markov Random Field (MRF) models (Atkinson and Lewis, 2000). However, such algorithms assume little or no prior knowledge of the scene; if some knowledge were available, then we should be able to increase the precision of prediction further. The approach described here is called per-parcel or per-field classification (Ortiz *et al.*, 1997; Aplin *et al.*, 1999) and was first proposed almost 20 years ago (Mason *et al.*, 1988).

4.4.2.1 Per-field classification

One of the problems with per-pixel classification is that, as described above, the pixel locations bear little or no relation to the scene. Per-parcel classification invokes the idea of a scene model in which the area of interest is comprised of objects (Johnsson, 1994; Mattikalli *et al.*, 1995). The vector model provides a bridge between the image and the object-based view of reality, which is often appropriate for scenes that have been developed by humans. A prominent example of the integration of remote sensing and GIS vector data for classification of land cover is the CLEVER mapping approach used to map land cover in the UK in 2000 (Smith and Fuller, 2001; Fuller *et al.*, 2002).

Per-parcel classification presupposes that the scene of interest can be represented as a series of objects arranged either (a) on a background (e.g. residential urban landscape in the UK) or (b) as a mosaic (e.g. agricultural landscape in the UK) (Strahler *et al.*, 1986). These models fit conceptually with our interpretation of urban and agricultural landscapes (the human brain has evolved to recognize functional objects; think of houses, roads, train stations) and so it is commonly these objects that we wish to label through remote sensing classification. Crucially, vector data help the investigator to separate the within-parcel variation (unwanted noise given the above) from the between-parcel variation (signal). Without such data, such separation is difficult to achieve. It is surprising, then, that per-parcel classification is not more widely adopted.

Integrating raster and vector data sources is replete with problems, most notably imprecision in geometric registration, as discussed above. Here, it is assumed that both datasets are perfectly registered in the same coordinate system. It is possible to label each pixel with a polygon ID using a GIS. Once each pixel is so labelled, it can be treated as part of a group, a member of a given parcel. Then, per-parcel classification can proceed in one of three basic ways: (a) averaging each pixel vector (remember, each pixel represents a set of wavebands) per-parcel prior to classification of each mean vector; (b) classification of the entire multivariate, multiwaveband dataset per-parcel; or (c) post-processing of the classified pixels per-parcel (e.g. taking the modal land cover class per-parcel). Option (c) is the simplest and most common approach (e.g. Aplin et al., 1999).

Per-parcel classification imparts benefits to the user beyond simple averaging over desired objects. For example, the vector data allow texture measures (e.g. the variogram) to be estimated per-object, removing the between-parcel variation that so often reduces the utility of such texture measures in remote sensing (Berberoglu et al., 2000; Lloyd et al., 2006). Further, it is possible to undertake a more sophisticated analysis of the distribution of classes allocated per-parcel [approach (c)]. For example, where the modal class corresponds to a large proportion (e.g. 95% of a parcel), one may be confident that the correct class has been allocated. Where the modal class is 45% and the second most common class represents 40%, then uncertainty is clearly greater. One interpretation of this particular situation is that there may be a missing line which should divide the parcel into two parts (such missing lines are common in agricultural scenarios, where farmers rotate crops, etc.) (Aplin and Atkinson, 2004). In these circumstances, the missing line can be added and incorrect allocation of the single class to the whole parcel avoided.

In per-field classification, vector data defining the objects of interest in the scene are combined with image data to allocate a single class to each object. The vector data are a representation or abstraction of the scene; they are a function of the scales of measurement (generalization) described in section 4.2. For this reason, the vector data constrain the outcome of the classification; specifically the geometry of each object being classified. In a sense, the vector data are taken as correct and the remotely sensed imagery is used to label these correct objects. Thus, the goodness of the classification very much depends on the level of generalization in the original vector data and, in particular, whether this level of generalization is appropriate for the classification task. Too generalized a representation may lead to misclassification of pixels near to the object boundaries.

4.4.2.2 Soft classification and subpixel allocation

In the preceding section it was assumed that classification involved hard allocation, in the sense that a *single* class was allocated to each pixel (or parcel). However, in remotely sensed images of scenes that are adequately conceptualized as comprising a set of objects, many pixels actually represent a mixture of classes; in particular, mixing occurs where pixels straddle the boundaries between two or more objects of different classes (Cracknell, 1998). In these circumstances, it makes little sense to allocate a single class to a pixel.

For the above reasons, soft classification approaches have become popular in remote sensing, in which a single pixel is allocated to multiple classes in proportion to the area covered by each class within the pixel (Foody, 1996). In many cases, the scale of measurement (pixel size) in the remotely sensed image is much finer than the size of the objects of interest (e.g. Landsat Thematic Mapper imagery with a spatial resolution of 30 m in relation to agricultural parcels of around 300 m on a side). In these circumstances, hard classification followed by subsequent per-parcel allocation through selection of the modal class may be adequate. However, in other cases the spatial resolution may be such that the pixel size interacts strongly with the objects, leading to much mixing within pixels [e.g. Système Pour l'Observation de la Terre (SPOT) multispectral imagery with a spatial resolution of 20 m in relation to a residential scene]. In these circumstances, per-parcel classification based on hard classification of pixels would be of limited use.

Aplin and Atkinson (2001) developed a procedure to deal with the above problem. Specifically, soft classification was used to allocate each pixel in a remotely sensed scene to multiple classes. Then, each class proportion within each pixel was allocated to one of the parcels that the pixel covered. This allocation was adjusted according to the proportion of the parcel that was covered (largest class proportions being allocated to the largest parcel proportions) (Figure 4.7). In this way, it was possible to gain the benefits of per-parcel classification even where the objects of interest were not much larger than a pixel. The utility of this procedure clearly depends on: (a) the scale of measurement relative to the underlying scale of variation (in this case, the size of the objects); and (b) accurate geometric registration between the vector and raster datasets.

4.4.2.3   A note on downscaling and super-resolution mapping

In recent years, several groups of researchers have developed techniques for subpixel or super-resolution mapping (Atkinson, 1997, 2004; Tatem *et al.*, 2001), in which the land cover proportions output from a soft classification are processed further to map the land cover *within* individual pixels. This amounts to downscaling – producing a land cover classification at a spatial resolution finer than that of the original imagery. Such a procedure depends on constrained optimization; specifically, the goal is to maximize the correlation between neighbouring subpixels

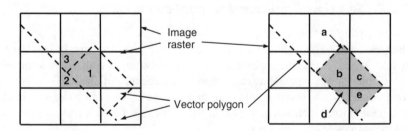

Figure 4.7   Calculating modal land cover per polygon by (a) assigning and (b) grouping sub-pixel land cover proportions

(pixels within pixels), while at the same time maintaining the land cover proportions predicted at the original pixel level. Essentially, the goal is spatial clustering. Tatem *et al.* (2001) developed a Hopfield neural network to achieve this optimization efficiently.

The HNN approach was extended subsequently to match the spatial correlation (and, thus, match also the scales of spatial variation) in some training image (rather than maximizing the spatial correlation) using a similar optimization algorithm. This pattern-matching algorithm allowed super-resolution mapping of objects that were smaller than a pixel. The spatial correlation was represented using the variogram (Tatem *et al.*, 2002), although the two-point histogram (Deutsch and Journel, 1998; Atkinson, 2004) is a preferable alternative.

This section has concentrated on land cover classification, primarily because this book is focused on raster and vector data integration. If the scene is conceptualized as comprising objects (requiring a vector data model), then labelling those objects is an appropriate task. Nevertheless, there are circumstances in which the objective may be to predict continua, and knowledge of (data on) objects in the form of vector data can increase the precision of prediction through data integration within a GIS. For example, geostatistical models are not suited for application to scenes that comprise objects. The ability to separate an image into component parts allows the RF model of geostatistics to be applied appropriately to within-class or within-object variation. This amounts to a non-stationary geostatistical model defined using vectors.

4.5 Conclusion

The key conclusions of this chapter can be summarized as follows:

- The scales of spatial variation in observed data are a function of the underlying spatial variation in the property of interest and spatial sampling processes.

- Models of the effect of the sampling framework upon observed spatial variation allow upscaling and downscaling of both observed data and models characterizing the spatial variation.

- The effects of sampling processes invoked through measurement are propagated when data are integrated within a GIS; thus, it is important to understand such processes if data integration is to be undertaken appropriately.

- Scales of measurement in vector data may introduce an effect through data integration that is similar, but spatially different, to the effect of geometric error.

- Data transformation often has the effect of introducing a new scale of measurement that is added to the scales of measurement inherent in the original sampling. Thus, particular care is needed where data are transformed prior to integration.

Acknowledgements

The editor is thanked for his patience while this chapter neared completion.

References

Amrhein, C. and Wong, D. (1996) Research on the MAUP: old wine in a new bottle or real breakthrough? *Geographical Systems* **3**, 73–76.

Aplin, P. and Atkinson, P. M. (2001) Sub-pixel land cover mapping for per-field classification. *International Journal of Remote Sensing* **22**, 2853–2858.

Aplin, P. and Atkinson, P. M. (2004) Predicting missing field boundaries to increase per-field classification accuracy. *Photogrammetric Engineering and Remote Sensing* **70**, 141–149.

Aplin, P., Atkinson, P. M. and Curran, P. J. (1997) Fine spatial resolution satellite sensors for the next decade. *International Journal of Remote Sensing* **18**, 3873–3881.

Aplin, P., Atkinson, P. M. and Curran, P. J. (1999) Fine spatial resolution satellite sensor imagery for land cover mapping in the United Kingdom. *Remote Sensing of Environment* **68**, 206–216.

Atkinson, P. M. (1997) Selecting the spatial resolution of airborne MSS imagery. *International Journal of Remote Sensing* **18**, 1903–1917.

Atkinson, P. M. (1999) Spatial statistics. In A. Stein and F. van der Meer (eds), *Spatial Statistics in Remote Sensing*. Kluwer: Dordrecht, The Netherlands, 57–81.

Atkinson, P. M. (2004) Resolution manipulation and sub-pixel mapping. In de Jong S. and F. van der Meer (eds), *Remote Sensing Image Analysis. Including the Spatial Domain*. Kluwer: Dordrecht, The Netherlands, 51–70.

Atkinson, P. M. (2004) Super-resolution land cover classification using the two-point histogram. In Sánchez-Vila X., Carrera J. and Gómez-Hernández J. (eds), *GeoENV IV: Geostatistics for Environmental Applications*. Kluwer, Dordrecht, The Netherlands, 15–28.

Atkinson, P. M., Foody, G. M., Wu, F. and Darby, S. E. (2005) *GeoDynamics*. CRC Press: Boca Raton, FL, USA.

Atkinson, P. M. and Curran, P. J. (1997) Choosing an appropriate spatial resolution for remote sensing investigations. *Photogrammetric Engineering and Remote Sensing* **63**, 1345–1351.

Atkinson, P. M., Dunn, R. and Harrison, A. R. (1996) Measurement error in reflectance data and its implications for regularizing the variogram. *International Journal of Remote Sensing* **17**, 3735–3750.

Atkinson, P. M. and Kelly, R. E. J. (1997) Scaling-up snow depth in the UK for comparison with SSM/I imagery. *International Journal of Remote Sensing* **18**, 437–443.

Atkinson, P. M. and Lewis, P. (2000) Geostatistical classification for remote sensing: an introduction, *Computers and Geosciences* **26**, 361–371.

Atkinson, P. M. and Martin, D. (1999) Investigating the effect of support size on population surface models. *Geographical and Environmental Modelling* **3**, 101–119.

Atkinson, P. M. and Tate, N. J. (2000) Spatial scale problems and geostatistical solutions: a review. *Professional Geographer* **52**, 607–623.

Berberoglu, S., Lloyd, C. D., Atkinson, P. M. and Curran, P. J. (2000) The integration of spectral and textural information using neural networks for land cover mapping in the Mediterranean. *Computers and Geosciences* **26**, 385–396.

Burrough, P. A. and McDonnell, R. A. (1998) *Principles of Geographical Information Systems*. Oxford University Press: Oxford, UK.

Chen, K. and Blong, R. (2003) Identifying the characteristic scale of scene variation in fine spatial resolution imagery with wavelet transform-based sub-image statistics. *International Journal of Remote Sensing* **24**, 1983–1989.

Chilès, J. P. and Delfiner, P. (1999) *Geostatistics: Modelling Uncertainty*. Wiley: New York, NY, USA.

Clark, I. (1977) Regularization of a semi-variogram, *Computers and Geosciences* **3**, 341–346.

Cracknell, A. P. (1998) Synergy in remote sensing – what's in a pixel? *International Journal of Remote Sensing* **19**, 2025–2047.

Csillag, F., Kertesz, M. and Kummert, A. (1996) Sampling and mapping of heterogeneous surfaces: multi-resolution tiling adjusted to spatial variability. *International Journal of Geographical Information Science* **10**, 851–875.

Curran, P. J., Atkinson, P. M., Milton, E. J. and Foody, G. M. (2000) Linking remote sensing, land cover and disease. *Advances in Parasitology* **47**, 37–80.

Curran, P. J. and Dungan, J. L. (1989) Estimation of signal-to-noise: a new procedure applied to AVIRIS data. *IEEE Transactions on Geoscience and Remote Sensing* **27**, 620–628.

Cushnie, J. L. (1987) The interactive effect of spatial resolution and degree of internal variability within land-cover types on classification accuracies. *International Journal of Remote Sensing* **8**, 15–29.

Deutsch, C. V., and Journel, A. G. (1998) *GSLIB: Geostatistical Software and User's Guide*, 2nd edn. Oxford University Press: New York, NY, USA.

Foody, G. M. (1996) Approaches for the production and evaluation of fuzzy land cover classifications from remotely-sensed data. *International Journal of Remote Sensing* **17**, 781–796.

Fotheringham, A. S., Brunsdon, C., Charlton, M. (2000) *Quantitative Geography: Perspectives on Spatial Data Analysis*. Sage: London, UK.

Fuller, R. M., Smith, G. M., Sanderson, J. M., Hill, R. A. and Thompson, A. G. (2002) The UK Land Cover Map 2000: construction of a parcel-based vector map from satellite images. *Cartographic Journal* **39**, 15–25.

Goovaerts, P. (1997) *Geostatistics for Natural Resources Evaluation*. Oxford University Press: New York, NY, USA.

Janssen, L. L. F. and Molenaar, M. (1995) Terrain objects, their dynamics and their monitoring by the integration of GIS and remote sensing. *IEEE Transactions on Geoscience and Remote Sensing* **33**, 749–758.

Janssen, L. L. F., Jaarsma, M. N. and Van der Linden, T. M. (1990) Integrating topographic data with remote sensing for land-cover classification. *Photogrammetric Engineering and Remote Sensing* **56**, 1503–1506.

Johnsson, K. (1994) Segment-based land-use classification from SPOT satellite data. *Photogrammetric Engineering and Remote Sensing* **60**: 47–53.

Journel, A. G. and Huijbregts, C. J. (1978) *Mining Geostatistics*. Academic Press: London, UK.

Jupp, D. L. B., Strahler, A. H. and Woodcock, C. E. (1988) Autocorrelation and regularization in digital images I. Basic theory. *IEEE Transactions on Geoscience and Remote Sensing* **26**, 463–473.

Jupp, D. L. B., Strahler, A. H. and Woodcock, C. E. (1989) Autocorrelation and regularization in digital images II. Simple image models. *IEEE Transactions on Geoscience and Remote Sensing* **27**, 247–258.

Lloyd, C. D., Berberoglu, S., Curran, P. J. and Atkinson, P. M. (2004) Per-field mapping of Mediterranean land cover: A comparison of texture measures. *International Journal of Remote Sensing* **15**, 3943–3965.

Mallat, S. G. (1989) A theory for multiresolution signal decomposition: the wavelet representation. *IEEE Transactions on Pattern Analysis and Machine Intelligence* **11**, 674–693.

Manslow, J. and Nixon, M. S. (2002) On the ambiguity induced by a remote sensor's PSF. In Foody G. M. and Atkinson P. M. (eds), *Uncertainty in Remote Sensing and GIS*. Wiley: Chichester, UK, 37–57.

Martin, D. (1996) An assessment of surface and zonal models of population. *International Journal of Geographical Information Systems* **10**, 973–989.

Mason, D. C., Corr, D. G., Cross, A. *et al.* (1988) The use of digital map data in the segmentation and classification of remotely-sensed images. *International Journal of Remote Sensing* **2**, 195–215.

Mattikalli, N. M., Devereux, B. J. and Richards, K. S. (1995) Integration of remotely sensed satellite images with a geographical information system. *Computers and Geosciences* **21**, 947–956.

McBratney, A. B. and Webster, R. (1986) Choosing functions for semi-variograms of soil properties and fitting them to sampling estimates. *Journal of Soil Science* **37**, 617–639.

Moellering, H. and Tobler, W. R. (1972) Geographical variances. *Geographical Analysis* **4**, 35–50.

Openshaw, S. (1984) The modifiable areal unit problem. *Concepts and Techniques in Modern Geography CATMOG* 38. Geo-Abstracts: Norwich, UK.

Ortiz, M. J., Formaggio, A. R., and Epiphanio, J. C. N. (1997) Classification of croplands through integration of remote sensing, GIS and historical database. *International Journal of Remote Sensing* **18**, 95–105.

Pebesma, E. J., Wesseling, C. G. (1998) Gstat, a program for geostatistical modelling, prediction and simulation. *Computers and Geosciences* **24**, 17–31.

Quattrochi, D. A. and Goodchild, M. F. (eds). (1997) *Scale in Remote Sensing and GIS*. CRC Press: New York, NY, USA.

Smith, G. M. and Fuller, R. M. (2001) An integrated approach to land cover classification: an example in the Island of Jersey. *International Journal of Remote Sensing* **22**, 3123–3142.

Strahler, A. H., Woodcock, C. E. and Smith, J. A. (1986) On the nature of models in remote sensing. *Remote Sensing of Environment* **20**, 121–139.

Tatem, A. J., Lewis, H. G., Atkinson, P. M. and Nixon, M. S. (2001) Super-resolution target identification from remotely sensed images using a Hopfield neural network. *IEEE Transactions on Geoscience and Remote Sensing* **39**, 781–796.

Tatem, A. J., Lewis, H. G., Atkinson, P. M., and Nixon, M. S. (2002) Super-resolution land cover pattern prediction using a Hopfield Neural Network. *Remote Sensing of Environment* **79**, 1–14.

Townshend, J. R. G. and Justice, C. O. (1988) Selecting the spatial resolution of satellite sensors required for global monitoring of land transformations. *International Journal of Remote Sensing* **9**, 187–236.

Tso, B. C. K. and Mather, P. M. (2001) *Classification Methods for Remotely Sensed Data*. Taylor and Francis: New York, NY, USA.

Ware, J. M. and Jones, C. B. (1998) Conflict reduction in map generalization using iterative improvement. *Geoinformatica* **2**: 383–407.

Webster, R. and Oliver, M. A. (1992) Sample adequately to estimate variograms of soil properties. *Journal of Soil Science* **43**: 177–192.

Westmoreland, S. and Stow, D. A. (1992) Category identification of changed land-use polygons in an integrated image processing/geographic information system. *Photogrammetric Engineering and Remote Sensing* **58**, 1593–1599.

Woodcock, C. E. and Strahler, A. H. (1987) The factor of scale in remote sensing. *Remote Sensing of Environment* **21**, 311–322.

5

Of patterns and processes: spatial metrics and geostatistics in urban analysis

XiaoHang Liu* and Martin Herold†

*Department of Geography and Human Environmental Studies, San Francisco State University, USA
† Institut für Geographie, Friedrich-Schiller-Universität Jena, Germany

5.1   Introduction

Population growth, combined with technology advancement, has led to continued urbanization at the global scale. To assess the social, economic and environmental impact of this phenomenon, an understanding of the urban patterns and processes is necessary. Answers to questions on how cities are spatially organized, where and when developments happen, and how different processes lead to different patterns can significantly enrich our understanding of the urban system and forward scientific planning. To this end, detailed spatial and temporal knowledge of an urban area is essential, including information about the area's morphology, infrastructure, land use pattern and population distribution and the drivers behind its growth. The potential of satellite remote sensing to provide such information has been widely discussed in the literature (Batty and Howes, 2001), especially with the recent availability of very high spatial resolution images, such as IKONOS and QUICKBIRD (Donnay *et al.*, 2001). For clarification, the study of *urban morphology* is concerned with how a city grows spatially and is directly linked to the architectural domain. (*Urban*) *pattern* more generally refers to the spatial heterogeneity of cities in the context of a variety of issues characterizing urban environments (social, economic, etc.). *Processes* describe the evolution and change of urban areas. Observed urban

Integration of GIS and Remote Sensing   Edited by Victor Mesev
© 2007 John Wiley & Sons, Ltd.

patterns and morphological characteristics are an outcome of the processes that formed them. Thus, both are directly linked and usually studied in conjunction.

While satellite remote sensing brings excitement to the field of urban analysis, challenges remain. One is that remote sensing records what is on the land surface, i.e. land cover information, whereas in urban analysis, it is *land use* that is of primary interest. Land use refers to the purpose served by a land. For example, a vegetated area can be of recreational land use if it is in an urban park, or of residential land use if it is the playground in a residential area. Remote sensing provides information at the *land cover* level. How to link land cover to land use remains a challenge. This challenge is especially prominent in urban areas because of the spatial and spectral heterogeneity of urban environments (Jensen and Cowen, 1999; Herold *et al.*, 2003a). Another critique of urban remote sensing is that, although remote sensing images provide vast amounts of spatial and temporal details of urban area, remote sensing is largely blind to patterns and processes (Longley, 2002). Without appropriate methods with which to distil patterns and link these patterns with the drivers and processes behind them, information in remote sensing images can only be marginally used in urban analysis. In this context, Longley and Tobon (2004) pointed out that development of direct, timely and spatially disaggregate urban indicators that can be derived from remote sensing images is key to a new data-rich and relevant urban analysis.

In this chapter, we introduce geostatistics and spatial metrics as two methods of addressing these challenges and illustrate their utility in three case studies: to link land cover to land use, to link urban form to population density, and to link urban pattern to the growth drivers. The functions of geostatistics and spatial metrics in these case studies are two-fold: as descriptors of image texture and as indicators of urban form. Image texture refers to the tonal change in the image. It is closely related to pattern, which is the spatial arrangement of textural components (Brivio and Zilioli, 2001). Texture is an important interpretation key in areal interpretation, together with shape, context, association, etc. (Haack *et al.*, 1997), because it describes the spatial variability of ground features. Considering that very high-resolution satellite images are almost comparable to aerial photographs, texture is also likely to help derive information from satellite images. It is in this context that we examine the utility of geostatistics and spatial metrics as texture descriptors to link land cover to land use. The other usage of geostatistics and spatial metrics is to describe urban form and pattern. Morphology and pattern are the products of development processes. As discovered by research in urban growth modelling, drivers behind the development processes can be social and economic as well as physical. Effective urban indicators should thus be able to provide insight on the linkage between urban patterns and processes.

Before introducing geostatistics and spatial metrics as two methods to describe image texture and the spatial variability in urban areas, it has to be pointed out that other methods are also available, such as the standard deviation method (Arai, 1993), the contrast between neighbouring pixels (Edwards *et al.*, 1988), local variance

(Woodcock and Harward, 1992), fractal dimensionality (Batty and Longley, 1994; De Jong and Burrough, 1995), grey-level co-occurrence matrix (Haralick, 1973) and wavelet analysis (Zhu and Yang, 1998). The studies vary in terms of image type, spatial unit and the statistical method. The geostatistics and spatial metrics methods to be introduced in this chapter are more complex. They are discussed as new approaches to analyse satellite images of very high spatial resolution.

5.2 Geostatistics

Initially developed in the field of mining, geostatistics is applied statistics based on theories of regionalized variables (Webster, 1973). Regionalized variables describe phenomena with spatial distribution and exhibit spatial continuity. In the context of remote sensing, the reflectance value of a pixel can be considered as a function of its geographic location, hence a realization of a regionalized variable (Gooverts, 1999). A key concept in geostatistics is spatial autocorrelation, meaning that observations that are close to each other in geographic space tend to have more similar values than observations far apart. Spatial autocorrelation suggests that the variance between observations in the near range tends to be small and increases with distance. The geostatistical tool to describe the scale and pattern of spatial variability is called semi-variance or the structure function, and is calculated by the following equation:

$$\hat{\gamma}(h) = \frac{1}{2N_h} \sum_i (z_i - z_{i+h})^2 \qquad (5.1)$$

where h is the lag distance along a specified direction, i.e. it is a vector, $\hat{r}(h)$ is the semi-variance at lag h, z_i and z_{i+h} are the values of a pair of pixels separated by h, and N_h is the number of such pairs in the study area. By varying the value of h and calculating the corresponding $\hat{r}(h)$ using equation 5.1, an experimental semi-variogram can be generated. Note that in using equation 5.1, a pixel is assumed to be a point. In reality, each pixel corresponds to a certain extent on the ground. This scale inconsistency is discussed by Lark (1996) and Jupp et al. (1998). In practical analysis, equation 5.1 is often used.

The application of geostatistics in remote sensing is abundant in the literature. Applications include image interpolation (Herzfeld, 1999), contextual classification (Kyriakidis et al., 2004) and uncertainty mapping (De Bruin, 2000). In the context of studying land use patterns, geostatistics can reveal the spatial variability of urban structures. For example, a comparison study of TM images of different cities reveals that a large metropolitan area has significantly higher semi-variance than that of a small city (Brivio and Zilioli, 2001). In this chapter, we examine the utility of semi-variances to describe the characteristics of different urban land uses, and the linkage of land use pattern to human population distribution. Unlike previous studies, the semi-variance analysis in this chapter is based on an image of very high spatial resolution, i.e. the 4 m multispectral IKONOS image. Such images are generally considered to have the highest potential for urban remote sensing (Jensen and Cohen, 1999).

The lag distance h in equation 5.1 is a vector, which means that it is associated with both a distance and a direction. When the ground features do not show any directional organization, i.e. the value of $\hat{r}(h)$ in equation 5.1 does not depend on the direction but distance only, h is reduced from a vector to a scalar. The semi-variogram obtained is considered isotropic (omni-directional); otherwise it is anisotropic. In this chapter, we illustrate geostatistics using isotropic variograms only. Another thing to note is that scale also has an effect on the value of semi-variograms. For the same study area, the experimental semi-variogram may vary depending on the spatial resolution of the remote sensing image. Additionally, the spatial unit selected to calculate a semi-variogram can also have an impact. In the literature, the spatial unit used to conduct semi-variogram analysis is usually a square window (e.g. 7×7 pixels) and little guidance is available on selecting the size of the window. In this study, we use land use zones as the spatial unit, which has a better correspondence with urban patterns.

5.3 Spatial metrics

The semi-variances method in equation 5.1 works with the digital values of pixels. The second method we introduce, spatial metrics, is rather different, in the sense that it works with categorical information. Spatial metrics were first developed in the field of landscape ecology to describe the composition of a natural landscape. The rationale behind using spatial metrics to describe the composition and pattern of an urban landscape is as follows. An urban area is characterized by diverse materials, such as concrete, asphalt, metal, plastic, glass, shingles, water, grass, shrubs, trees and soil (Jensen and Cowen, 1999). Although the reflectance values of these materials are different, they can be grouped into a few land cover categories, such as vegetation, built-up area, etc. In fact, Ridd (1995) proposed a V–I–S model to categorize urban areas, where V, I and S stand for vegetation, impervious surface and soil, respectively. The land use of an area depends on the proportion and the spatial configuration of these land covers (Figure 5.1). Spatial metrics have the capability to describe the composition and spatial arrangement of the land covers in a landscape. Therefore, they can be used to describe urban patterns and structures. When applying spatial metrics, the spatial unit used is called a patch. A patch is defined as an object made up of pixels which are adjacent to each other and have the same land cover. For example, a house occupying 20×20 m on the ground might correspond to 5×5 pixels in a multispectral IKONOS image. Although these 5×5 pixels may have different spectral reflectance values in the remote sensing image, their land covers are all classified as built-up, hence forming a continuous patch. Similarly, a contiguous vegetated area such as a lawn also forms a patch.

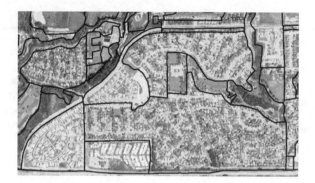

Figure 5.1 Examples of land use zones in the study area

Many spatial metrics have been developed. Table 5.1 provides a list of those spatial metrics that are especially related to composition and spatial configuration. A thorough discussion of spatial metrics and their linkage to remote sensing is provided McGarigal *et al.* (2002) and Herold *et al.* (2003a).

Although spatial metrics are relatively new to the urban analysis community, their usage is increasing rapidly. Currently, spatial metrics have been used on urban land use/land cover classification (Herold *et al.*, 2003b) and population density estimation (Liu *et al.*, 2005). Their usage in urban modelling is especially substantive. Herold *et al.* (2005) documented the role of spatial metrics in the analysis and modelling of land use change. It is argued that urban landscape composition and pattern, as described by spatial metrics, are critical independent measures of the economic landscape function and can therefore provide an improved interpretation and evaluation of modelling results. Parker *et al.* (2001) summarized the usefulness of spatial metrics with respect to a variety of urban models and argued for the contribution of spatial metrics in helping link economic processes and patterns of land use. Alberti and Waddell (2000) substantiated the importance of spatial metrics in urban modelling by using them to model the effects of complex spatial patterns of urban land cover and land use on social and ecological processes. Geoghegan *et al.* (1997) explored spatial metrics in modelling land and housing values. They found that housing price is influenced by the land use patterns surrounding a parcel, indicating that people care very much about the landscapes around them. The authors recommended spatial metrics to describe such relationships. Recently, Dietzel *et al.* (2005) illustrated how spatial metrics can be used to link the empirical evidence of spatio-temporal dynamics of land use to urban theory. In this chapter, we demonstrate the utility of spatial metrics in linking land cover to land use, and linking urban form to other urban characteristics, such as population distribution and urban growth factors.

One challenge in using spatial metrics is how to choose the most effective ones. Currently, little guidance exists on spatial metric selection, since the suitability of

Table 5.1 Examples of the spatial metrics

Metrics description	Description
PLAND (percentage of landscape) $$PLAND = P_k = \frac{\sum_l a_{kl}}{A}(100)$$ $P_k =$ proportion of the landscape occupied by land-cover class k $a_{kl} =$ area (m^2) of patch kl, i.e. patch l with land cover k $A =$ total landscape area (m^2)	PLAND quantifies the proportional abundance of each patch type in the landscape. For this research, it is used to describe the percentage of buildings, vegetation, and other land cover types in a land use polygon
PD (patch density) $$PD = \frac{N}{A}(10,000)(100)$$ $N =$ total number of patches in the landscape $A =$ total landscape area (m^2)	PD describes the number of patches on a per unit area basis that facilitates comparisons among land use polygons of varying size. The higher the PD, the more fragmented the land use polygon. High-density single-unit housing is expected to have the highest PD value, while forest/rangeland may have the lowest
LPI (largest patch index) $$LPI = \frac{\max(a_{ij})}{A}(100)$$ $a_{kl} =$ area (m^2) of patch kl, i.e. patch l with land cover k $A =$ total landscape area (m^2)	LPI quantifies the dominance of the largest patch. A fragmented landscape (e.g. high-density single-unit housing) has a lower LPI value than a less fragmented landscape (e.g. agricultural land)
ENN (Euclidean nearest-neighbour distance) $ENN = h_{kl}$ $h_{kl} =$ distance (m) from patch kl to the nearest neighbouring patch of the same class k, based on patch edge-to-edge distance, computed from cell centre to cell centre ENN-MN, Euclidean mean nearest-neighbour distance; ENN-SD, Euclidean nearest-neighbour distance SD	ENN equals the distance (m) to the nearest neighbouring patch of the same type, based on shortest edge-to-edge distance. ENN is a simple measure of isolation. For buildings, ENN can help to describe whether the houses are spaced regularly. High-density single-unit housing displays the highest orderliness and closeness to each other. Therefore, its ENN is expected to be small and has a low SD
AREA (patch area) $$AREA = a_{kl}\left(\frac{1}{10,000}\right)$$	AREA_MN and AREA_SD describes the uniformity of the patches comprising a landscape mosaic

$a_{ij} =$ area (m^2) of patch kl, i.e. patch l with land cover k

AREA_MN, mean patch areas in the landscape; AREA_SD, SD of patch areas in the landscape

ED (edge density)

$$ED = \frac{\sum\limits_{k=1}^{m} e_{ik}}{A}(10,000)$$

$e_{ik} =$ total length (m) of edge in landscape involving land cover type i

$A =$ total landscape area (m^2)

ED measures the total edge length of a landscape on a per unit area basis. For landscapes of similar size, the more it is fragmented, the higher is its edge density

FRAC_AM (area-weighted mean patch fractal dimension)

$$FRAC - AM = \sum_{j=1}^{n} \left[\frac{2\ln(.25p_{ij})}{\ln a_{ij}} \cdot \left(\frac{a_{ij}}{\sum\limits_{j=1}^{n} a_{ij}} \right) \right]$$

p_{ij}, perimeter (m) of patch ij; a_{ij}, area (m^2) of patch ij

FRAC_AM describes the shape complexity of a landscape. Its value is between 1 and 2. FRAC_AM approaches 1 for landscapes with simple shape of perimeter. FRAC_AM approaches 2 for highly complex shapes of perimeters. Fractals have been used to link urban form and functions (Batty and Longley, 1994)

$$Cohesion = \left[1 - \frac{\sum\limits_{j} p_{kl}^*}{\sum\limits_{l} p_{kl}^* \sqrt{a_{kl}^*}} \right] \cdot \left[1 - \frac{1}{\sqrt{z}} \right]^{-1} \cdot (100)$$

p_{kl}^*, perimeter of patch kl in terms of number of cell surfaces; a_{kl}^*, area of patch kl in terms of number of cells; Z, total number of cells in the landscape

Patch cohesion index measures the physical connectedness of the corresponding patch type. Cohesion approaches 0 as the proportion of the landscape comprised of the focal class decreases and becomes increasingly subdivided and less physically connected. Cohesion increases monotonically as the proportion of the landscape comprised of the focal class increases

---

Modified after McGarigal et al. (2002).

specific metrics depends on the objective of the study and the landscape characteristics of the study area. Also, in the context of spatial metrics and remote sensing, the categorical map on which spatial metrics are applied is usually generated by interpreting remote sensing images. This means any change in image resolution and thematic and semantic class definitions can affect the accuracy of the land cover map and hence the performance of the spatial metrics.

5.4   Examples

5.4.1   Data preparation

In this section, we illustrate the utility of geostatistics and spatial metrics for urban analysis through three examples: the linkage between land cover and land use, the linkage between urban morphology and population distribution, and the linkage between urban patterns and growth processes. All three examples are based on a dataset compiled for Santa Barbara, CA, USA. The remote sensing image used is a mosaic of seven individual IKONOS images acquired during March–July 2001. Since the images were acquired on different dates with varying atmospheric and illumination conditions, geometric and atmospheric corrections were conducted using standard image analysis algorithms to create a geometrically rectified and normalized image mosaic. Details on the preprocessing of the IKONOS images can be found in Herold *et al.* (2002b). For this study, the four multispectral bands with a spatial resolution of 4 m were used.

The IKONOS image was digitized into land use zones by an experienced remote sensing specialist. Each land use zone was assigned to one of the nine land use categories in Table 5.2 through visual interpretation. A land use zone is a photomorphic region (Peplis, 1974), which refers to an image segment with a homogenous image texture visibly different from that of the neighbouring land use zones. In residential areas, the structures of the built-up areas within a land use zone are similar in terms of size, density and spatial pattern. The boundary of a land use zone follows streets and other relevant natural and anthropogenic features whenever possible. Figure 5.1 is an example of the land use zones in the IKONOS image.

The IKONOS image is classified into three land cover types: buildings, green vegetation, and other, which includes roads, parking lots, bare soil, water bodies and non-photosynthetic vegetation. The rationale for these three land cover types is the V–I–S (vegetation–impervious surface–soil) model proposed by Ridd (1995), who demonstrated that different urban land uses can be characterized by the composition of vegetation, impervious surface and bare soil. To identify the three land cover types in the IKONOS image, an object-orientated approach is utilized using e-cognition (Herold *et al.*, 2002). Green vegetation spectrally separates fairly well. However, there are some spectral similarities between buildings or roof types and other urban targets, such as roads and bare soil surfaces, especially given the relatively low spectral resolution of IKONOS. The output is a land cover map with each pixel labelled as building, vegetation or other. Contiguous pixels with the same land-cover type are then aggregated to form land cover patches which are to be used by spatial metrics. The overall accuracy of the land cover map is assessed as 82.4%, and the kappa coefficient is 71.4%. Green vegetation is mapped with the highest accuracy, with a tendency to be overmapped. There is some confusion between the buildings/roofs and the other land-cover classes, due to the aforementioned spectral

Table 5.2 Land use classification based on semi-variogram

Class		Producer accuracy	No. of samples	LSU 1	MSU 2	HSU 3	MU 4	INST 5	REC 6	CI 7	AgR 8	FW 9
LSU	1	53.9	39	21	5	5	–	2	2	4	–	–
MSU	2	56.7	30	2	17	4	5	–	–	2	–	–
HSU	3	75.6	41	–	–	31	9	–	–	1	–	–
MU	4	76.5	34	–	3	4	26	–	–	1	–	–
INST	5	63.6	22	4	1	–	2	14	–	1	–	–
REC	6	54.5	22	–	–	–	–	1	12	4	4	1
CI	7	52.3	44	–	4	3	5	1	2	23	4	2
AgR	8	54.6	44	1	–	2	–	1	6	6	24	4
FW	9	35.7	28	–	–	–	–	5	8	5	–	10
Total			304	28	30	49	47	24	30	47	32	17
User accuracy				75.0	56.7	63.3	55.3	58.3	40.0	48.9	75.0	58.8

Overall accuracy $= 58.6\%$; $\kappa = 53.1\%$.

similarity. These errors in the categorical land cover map could be propagated to subsequent analysis using spatial metrics.

Nine land uses were identified in the study area. These land use types vary in terms of land cover composition and spatial variability. They are: residential uses, including low-density single-unit (LSU) area, medium-density single-unit area (MSU), high-density single unit area (HSU), multiple unit area (MU); commercial and industrial land use (CI); institutional land use (INST); recreational and open space (REC); agriculture (AgR); and forest and wetlands (FW).

The definitions of these land use types are found in Table 5.2. These nine land use types are commonly found in urban areas, although the definitions may vary from city to city. Each land use zone digitized from the IKONOS image is assigned to a land use category through visual interpretation by the image analyst. Figure 5.2 illustrates the different land uses in the IKONOS image and the classified land cover map. For illustration purposes, a map showing the accurate shape and location of building footprints and roads is included in Figure 5.2. By comparing the IKONOS land-cover classification map with the map of building footprints and roads, it can be seen that the land-cover classification is fairly accurate.

In contrast to spatial metrics, which use categorical land cover information, a semi-variogram is applied to grey-level images. Several grey-level images were experimented upon, including each of the four bands of the multispectral IKONOS image and a grey-level image based on the normalized difference vegetation index (NDVI) of each pixel. NDVI was chosen because it can efficiently differentiate built-up areas from vegetation. The NDVI image was used for the semi-variogram analysis in this chapter.

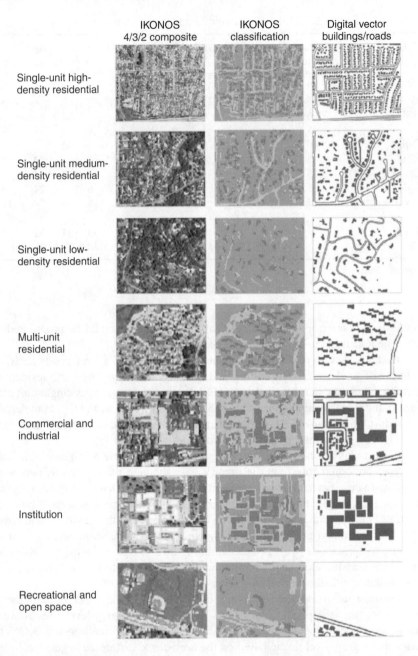

Figure 5.2 Examples of spatial land cover configurations for major urban land-use categories shown as an IKONOS false-colour composite, and an IKONOS classification result with buildings in red and vegetation in green. The digital vector building/roads data are included here to provide a ground reference of built-up area

5.4.2 Linkage from land cover to land use

5.4.2.1 Land use classification based on geostatistics

A semi-variogram was calculated for each land use zone using the NDVI image. In calculating the semi-variogram value, directions are assumed to be isotopic and the lag distances varied in the range 1–30 pixels. Figure 5.3 portrays how semi-variances change with different land use classes. It can be seen that vegetated areas (forest and agriculture) have the lowest variances, followed by recreational/open space and commercial land use. Residential areas display higher variances, and are very similar to each other. For residential classes and commercial land use, the semi-variograms reach the sill on or before lag 30, while for other classes, the semi-variance is still increasing. The semi-variogram profile in Figure 5.3 suggests that a lag distance of 30 is sufficient to differentiate the land use classes.

The similar semi-variograms among residential land uses suggest that it is not easy to use semi-variograms to differentiate these classes. To confirm this hypothesis, a classification exercise was conducted; 433 land zones were sampled from the study area, 129 zones were used to train the different classifiers, while the other 304 were reserved for accuracy assessment. Recall that the land use category of each land zone is assigned by a remote sensing expert through visual interpretation. The output is considered to be of such high accuracy that it can serve as the ground truth. Yet due to the exhaustive and exclusive classification scheme used, some land use zones fall between two categories. This is especially prominent in residential areas, where the boundary between low, medium and high

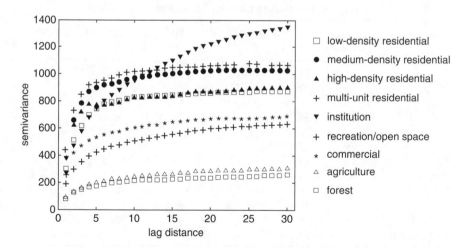

Figure 5.3 Plot of semi-variance of different land use classes using NDVI data

density is usually blurry. When such a 'transitory' zone is used to train or test accuracy, it is difficult to tell whether the error is due to the crisp classification scheme or the classifier utilized. A discussion of the problems associated with crisp classification can be found in Wang (1990). In this chapter, because our interest is the accuracy of a classifier, only the most representative zones in each land use category were sampled, thus resulting in 433 land use zones in total.

The semi-variograms of the training zones from each land use category were calculated. Their average was considered as the characteristic semi-variogram of that land use type. The semi-variograms of the test land use zones were compared with those of the training zones. A minimum distance classifier was used to identify the land use category of the test land use zones. Table 5.3 shows the result of this exercise. The overall accuracy is about 58% and the kappa coefficient is 52%. The confusion mainly occurs within two clusters, the residential cluster and the non-residential cluster, as indicated by Figure 5.3.

5.4.2.2 Land use classification based on spatial metrics

Like variograms, spatial metrics can also be used to describe land use characteristics. The spatial metrics method was applied to the land cover map, which consists of three classes: built-up, vegetation and other. Each land use zone is described using a vector of nine spatial metrics:

$$X = [PLAND_{built}, PLAND_{veg}, PD_{built}, PD_{veg}, ENN\_MN_{built}, ENN\_SD_{built},$$

$$COHESION, CONTAG, SHDI]$$

Table 5.3 Confusion matrix of metrics-based land use classification

Class		Producer accuracy	No. of samples	LSU 1	MSU 2	HSU 3	MU 4	INST 5	REC 6	CI 7	AgR 8	FW 9
LSU	1	48.7	39	19	10	–	–	–	8	1	1	–
MSU	2	73.3	30	1	22	–	5	–	–	2	–	–
HSU	3	70.7	41	–	–	29	11	–	–	1	–	–
MU	4	70.6	34	–	–	3	24	5	–	2	–	–
INST	5	72.7	22	–	1	0	1	16	–	4	–	–
REC	6	59.1	22	1	–	–	–	–	13	1	5	2
CI	7	70.5	44	–	2	2	3	6	–	31	–	–
AgR	8	47.7	44	3	2	–	1	–	11	–	21	6
FW	9	42.9	28	–	–	–	–	–	6	–	10	12
Total			304	24	37	34	45	27	38	42	37	20
User accuracy				79.2	59.5	85.3	53.3	59.3	34.2	73.8	56.8	60.0	

Overall accuracy = 61.5%; κ = 56.6%.

Figure 5.4 shows a comparison between different land uses described by spatial metrics. As in the experiment with semi-variograms, the spatial metrics of training and test land use zones were respectively calculated. A minimum distance classifier was then used to determine the land use category of each test zone. The results are listed in Table 5.6. Compared to the geostatistics method, spatial metrics performed slightly better, with an overall accuracy of 61.5% and kappa coefficient of 57.2%.

Compared to semi-variance, the meaning of the spatial metrics is more intuitive. For percentage of landscape (PLAND), it can be seen that the percentage of built-up areas increases as the land-use class moves from low-density residential to medium-density residential and high-density residential. Commercial areas also have a high percentage of built-up area but, compared to residential land uses, their percentage of vegetation is lower. In forest and agricultural areas, built-up land cover is barely present. Lo (1995) pointed out that vegetation is a good indicator of the population density of a neighbourhood. The plot of PLAND seems to support this statement.

Patch density measures the number of patches on a per unit area basis; the higher the patch density, the higher the degree of fragmentation. Houses in low-residential-density areas are usually far apart from each other. The vegetated areas separating two adjacent houses tend to be big. As the population density increases, the size of detached houses decreases, as does the vegetated area separating them. This suggests that as the residential population density increases, the patch density of both built-up and vegetation areas will decrease, as demonstrated by Figure 5.4b. MeanENN and stdENN respectively measure how far apart two adjacent houses are and the variation of that distance. Houses in low-residential-density areas display less orderliness and tend to be farther away from each other. It is therefore no surprise to see that meanENN and stdENN are both higher for low-density residential areas than for high-density residential areas. 'Contagion' measures how contagious a general patch is in the study area. Since built-up and vegetated patches in low-density areas tend to be big, this means that each patch in these areas is more 'contagious', therefore it has a higher contagion value than high-density residential areas. Shannon's diversity index is also correlated with the degree of fragmentation. Since high density areas are more fragmented, they have more patches per unit area, thus displaying a higher degree of diversity. The plots in Figure 5.4 show that all metrics seem to contribute to the differentiation of land-use classes. However, the cohesion values of all classes are over 90%. The difference is relatively low, suggesting that cohesion may not be an efficient feature for land-use classification.

5.4.2.3 Land-use classification based on combined information

A comparison between Tables 5.3 and 5.4 suggests that the overall semi-variance and kappa coefficients of spatial-metrics-based classification are similar. However,

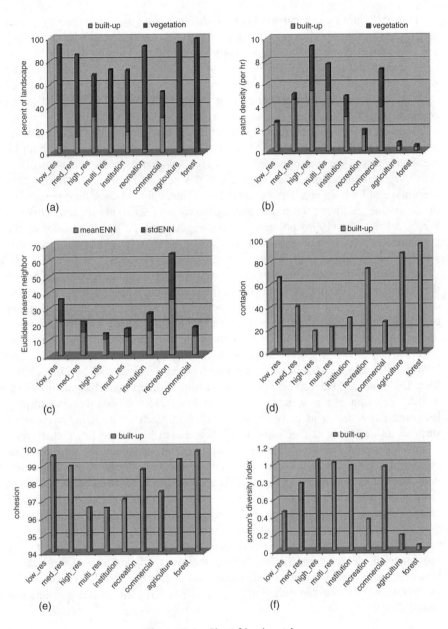

Figure 5.4 Plot of land metrics

Table 5.4 Land use differentiation based on semi-variogram and spatial metrics

Class		Producer accuracy	No. of samples	LSU 1	MSU 2	HSU 3	MU 4	INST 5	REC 6	CI 7	AgR 8	FW 9
LSU	1	69.2	39	27	8	–	–	–	4	–	–	–
MSU	2	76.7	30	3	23	2	2	–	–	–	–	–
HSU	3	61.0	41	–	1	25	14	–	–	1	–	–
MU	4	76.5	34	–	1	4	26	3	–	–	–	–
INST	5	77.3	22	–	–	–	4	17	–	1	–	–
REC	6	50.0	22	2	–	–	–	–	11	–	7	2
CI	7	70.5	44	–	–	2	4	7	–	31	–	–
AgR	8	59.1	44	3	1	–	–	3	6	–	26	5
FW	9	60.7	28	2	–	–	–	–	1	–	8	17
Total			304	37	34	33	50	30	22	33	41	24
User accuracy			73.0	67.7	75.8	52.0	56.7	50.0	93.9	63.4	70.8	

Overall accuracy $= 66.8\%$; $\kappa = 62.5\%$.

Table 5.5 Correlation between population density and urban form descriptors

Method	Descriptors of urban form	R^2
Semi-variance	var_i, $i = 1, \ldots, 20$; semi-variances with lags of 1–20 pixels	0.20
Spatial metrics	Percentage of built-up area (PLAND$_1$)	0.55
	Percentage of vegetated area (PLAND$_2$)	
	Patch density of built-up area(PD$_1$)	

their accuracy in individual land use classes varies. This suggests the potential to combine the two methods together to achieve a better land use differentiation. Table 5.5 lists the land use classification accuracy when the two methods were combined. It can be seen that the accuracy reached over 62%, which is about 10% higher than that using either method alone. The producer's and user's accuracies of residential classes improved significantly, except for multi-unit residential areas. The accuracies for recreation, agriculture and forest also improved, but their accuracies were not as high as that of commercial and some residential areas. All three classes are characterized by large patches of vegetated area, which is usually better discriminated using spectral information.

5.4.3 Linking urban form to population density

In this section, we examine the linkage between urban form and population density. Population density is important ancillary information from which to derive land use information (Mesev, 1998). Remote sensing has long been used to estimate urban population and socio-economic parameters. Example surrogates derivable from remote sensing images include the extent of an urbanized area, the image spectral reflectance value, the proportion of each land use class, etc. In this study, we use semi-variances and spatial metrics as descriptors of urban form and land use pattern to examine their linkage to residential population density. The rationale behind this linkage is that populations living in areas with similar housing characteristics tend to have similar population density. Housing characteristics are determined by the size of the houses, the greenness and other conditions. The interaction of these characteristics forms different patterns on remote sensing images. Semi-variance and spatial metrics are both capable of describing the spatial variability of an area. Therefore, they have the potential to link urban form to population density. 1578 census blocks with homogenous residential land use were collected in the study area. Their population densities were obtained from US Census 2000. The semi-variance and spatial metrics of each block were calculated and correlated with the natural logarithms of population density. Semi-variances corresponding to lag distance between 1 and 20 pixels were examined on the NDVI image. The R^2 obtained is about 0.20.

For the spatial metrics method, although nine spatial metrics were examined in land use classification, only three of them were found significant in correlation with population density: the percentage of built-up area, percentage of vegetation in the area, and the patch density of the built-up area. The linear correlation has the following form (Figure 5.5):

$$\ln(d) = 8.819 + 1.772p_1 - 2.612p_2 + 0.0632p_3, \ R^2 = 0.55 \qquad (5.2)$$

where d is the population density (people/ km^2); p_1 the percentage of built-up area (PLANDb); p_2 the percentage of vegetation (PLANDv); and p_3 the patch density of the built-up area (PDb).

The R^2 obtained is 0.55, which is significantly higher than that of the semi-variance method. The number 0.55 suggests two things. One is that there is indeed some correlation between urban form and population density, but the correlation is not high enough to make reliable estimates of population. The significant residuals suggest that although urban form can explain a significant amount of variance in population density, factors other than urban form also deserve examination. Another conclusion is that even though spatial metrics and semi-variances are both descriptors of urban form, spatial metrics clearly have a higher correlation with population density. This underscores Longley and Tobon's (2004) point, that appropriate urban indicators are key for urban analysis based on remote sensing. Clearly, when the appropriate indicators are used, more information can be inferred from remote sensing data.

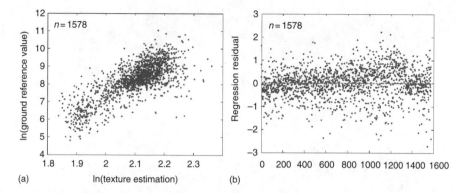

Figure 5.5 The correlation between spatial metrics and population density: (a) the comparison between texture-estimated population density and ground reference; (b) plot of the residual of the regression

5.4.5 Linking characteristics of spatial patterns and processes

The linkage between land use and population density suggests that land use is just one of the factors determining the spatial distribution of population in an area. The significant residuals left in equation 5.2 suggest that other factors may help to explain the variation in population distribution. Research in urban modelling suggests that the current population distribution pattern is the outcome of a development process. Such a process is often controlled and constrained by some spatial growth factors in the first place. In this section, we explore the linkage between urban form and growth factors, using the geographically weighted regression (GWR) technique (Fotheringham *et al.*, 2002). Unlike the linear regression technique used in section 4.5, which assumes that the relationship holds everywhere in the study area, GWR considers spatial non-stationarity and allows the regression relationship to vary over space. Specifically, given a location **u** in space, its local regression model can be written as:

$$y(\mathbf{u}) = \beta_0(\mathbf{u}) + \beta_1(\mathbf{u})x_1 + \ldots + \beta_n(\mathbf{u})x_n \qquad (5.3)$$

To estimate the values of the coefficients and the dependent variable at location **u**, a geographically weighting scheme is applied to fitting by least squares. The weighting scheme is organized such that data near location **u** are given a heavier weight in the regression than data further away. The result is that the regression determinant varies for different locations (Figure 5.6).

The analysis in section 5.4.5 shows that when urban forms are measured by spatial metrics, there is a stronger correlation between urban form and population density than if geostatistics are used. Based on this finding, in this section, the relationship between urban form and growth factors is examined using the spatial metrics method

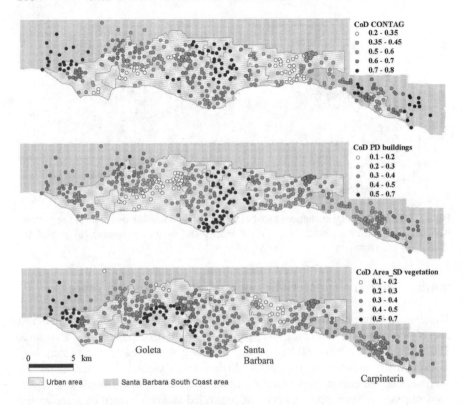

Figure 5.6 Spatial distribution of regression results for three growth factors vs. several spatial metrics describing residential urban pattern

only. Previous research has built an urban growth model for the study area (Herold *et al.*, 2003c) and found that topographic slope, distance to highways and distance to the central urban core are three very important factors controlling the population distribution in the area (Liu and Clarke, 2002). These three factors become the independent variables in GWR. Urban form, which is measured by spatial metrics, is the dependant variable. GWR is applied to examine to what extent the variances in the urban form can be explained by the three growth factors and how the power of this explanation varies over space. The spatial units of this analysis are the land use zones used in section 5.4.2 and illustrated in Figure 5.1. In describing the form of the land use zones, five spatial metrics were used: landscape contagion (CONTAG); percentage of buildings ($PLAND_b$); patch density of buildings (PD_b); patch density of vegetation (PD_v); and standard deviation of the areas of vegetation patches ($AREA\_SD_v$). The topographic slope of a land use zone is the average slope of the zone. In implementing GWR on a land use zone whose centroid is **u**, the distances to highways and to regional urban centres are calculated as that

Table 5.6 Results of the GWR analysis for multivariate regression models

	Landscape	Buildings		Vegetation	
	CONTAG	PLAND$_b$	PD$_b$	PD$_v$	AREA_SD$_v$
Local sample size	70	94	88	88	38
Global CoD (R^2)	0.55	0.31	0.34	0.31	0.34
T-slope	7.12	−6.39	−6.62	−5.94	0.13
T-core	7.91	−2.06	−3.8	−4.44	7.89
T-highway	8.05	−4.27	−4.02	−3.21	7.33
GWR CoD	0.73	0.47	0.51	0.47	0.69

Three growth factors are considered (distance to urban core, distance to highways, and slope) vs. five spatial metrics. One metric represents the heterogeneity of the landscape (CONTAG), two metrics describe characteristics of the land cover class building, two the class vegetation

of the centroid. A predefined number of nearest neighbours of **u** is searched and used in regression analysis. For example, in examining the correlation between CONTAG and the three growth factors, 70 nearest neighbours were used. Similarly, in running the correlation using PLAND$_b$, 94 nearest neighbours were used. The number of nearest neighbours can be specified or determined by the GWR software (Fotheringham *et al.*, 2002).

The result of GWR is summarized in Table 5.6. The value *Global CoD* (global coefficient of determination) can be interpreted as the R^2 of the global regression model based on all observations. GWR CoD, on the other hand, represents the R^2 of the geographically weighted regression. It provides a summary of the varying coefficients of determinants associated with different land use zones. The significance of each predictor in GWR is reflected in the *t*-values. *T-slope*, *T-core*, and *T-highway* refer to the *t*-value associated with slope, distance to urban centre and distance to highway, respectively. If the absolute *t*-value is above 1.96, the predictor is considered significant and retained for the regression model. *t*-Values below 0 indicate a negative linear relationship between the growth factor and the spatial metric. It can be seen from Table 5.6 that all three factors are significant in explaining the variances in urban form, except the standard deviation (SD) of the areas of vegetation patches (Area_SD$_v$).

A comparison between the Global CoD and GWR CoD suggests the improvements in the regression models using geographically weighted regression. This is expected, since local regression models make more adjustments to specific local characteristics than do global approaches. The overall GWR CoD for these metrics is in the range 46–72% (Table 5.2), meaning that the three growth factors are able to explain a significant proportion of the urban patterns which emerged as the result of historical growth processes. For each land use zone, its local CoD is associated with its centroid and mapped in Figure 5.6. It can be seen that the local CoD varies in different parts of the study area. In some areas, the relationship is more determined than in others. The spatial distribution of CoD of contagion (CONTAG) and patch density of developed area (PD_b) are fairly similar. Both show high correlations with the growth factors in residential areas and display low CoD in areas of agricultural land use, at urban edges and in the transition zones between urban centres. The morphology of these areas is better correlated with $AREA\_SD_v$, which describes vegetation pattern. From an urban development perspective, the areas which are highly correlated with CONTAG and PD_b are the residential areas which have been subjected to planning. The areas poorly correlated with CONTAG and PD_b are mostly unplanned or rural. These are usually the areas under development pressure, hence they are of more interest to urban planning. The boundaries of these areas are likely be changed by actions such as farmland protection or the establishment of urban growth boundaries.

5.5   Conclusion

The availability of a new generation of satellite images of very high spatial resolution suggests an unprecedented opportunity for utilizing remote sensing to study urban patterns and processes and the link between them. The vast amounts of spatial and temporal details provided by these images must be summarized into effective urban indicators in order to become useful and successful. It is in this context that the development of new image analysis methods is widely considered to be a pressing challenge. In this chapter, we introduced geostatistics and spatial metrics as two methods with which to derive urban information from remote sensing images. To illustrate their usage, three applications were presented: the usage of geostatistics and spatial metrics as image texture descriptors to infer urban land use information; the usage of geostatistics and spatial metrics as descriptors of urban form; and their linkage with other characteristics of the urban area such as population distribution and urban growth factors. Although both geostatistics and spatial metrics are valid descriptors of urban patterns, our study shows that the amount of urban information that can be derived from them are rather different. This suggests that their combined application provides a more comprehensive description and thus is more powerful than either descriptor alone. Clearly, the examples have shown potential, but much research remains to be conducted if remote sensing images are to be used to foster our understanding of urban areas from space and link empirical evidence to urban theory and practices.

References

Alberti M., and Waddell P. (2000) An integrated urban development and ecological simulation model. *Integrated Assessment* **1**: 215–227.

Arai, K. (1993) A classification method with a spatial-spectral variability. *International Journal of Remote Sensing* **14**, 699–709.

Batty, M. and Longley P. A. (1994) *Fractal Cities: A Geometry of Form and Function.* Academic Press: London, UK.

Batty, M. and Howes D. (2001) Predicting temporal patterns in urban development from remote imagery. In Donnay, J. P., Barnsley, M. J. and Longley P. A. (eds), *Remote Sensing and Urban Analysis.* Taylor and Francis: London, UK, 185–204.

Brivio, P. and Zilioli, E. (2001) Urban pattern characterization through geostatistical analysis of satellite images. In Donny, J. P., Barnsley, M. J., and Longley P. A. (eds), *Remote Sensing and Urban Analysis.* Taylor and Francis: London, UK, 43–60.

De Bruin, S. (2000) Predicting the areal extent of land-cover types using classified imagery and geostatistics. *Remote Sensing of Environment* **74**, 387–396.

De Jong, S. M. and Burrough P. A. (1995) A fractal approach to the classification of Mediterranean vegetation types in remotely sensed images. *Photogrammetric Engineering and Remote Sensing* **61**, 1041–1053.

Deutsch, C. and Journel, A. (1997) *GSLIB: Geostatistical Software Library and User's Guide*, 2nd edn. Oxford University Press: New York, NY, USA.

Dietzel, C., Herold, M., Hemphill, J. J. and Clarke, K. C. (2005) Spatio-temporal dynamics in California's Central Valley: empirical links urban theory. *International Journal of Geographic Information Sciences* **19**(2): 175–195.

Donnay, J. P., Barnsley, M. J. and Longley, P. A. (2001) Remote sensing and urban analysis. In Donnay, J. P., Barnsley, M. J. and Longley P. A. (eds). *Remote Sensing and Urban Analysis.* Taylor and Francis: London, UK, 3–18.

Edwards, G., Landary, R., and Thomson K. P. B. (1988) Texture analysis of forest regeneration sites in high-resolution SAR imagery. Proceedings of the International Geosciences and Remote Sensing Symposium (IGARSS 88), ESA SP-284, Paris, European Space Agency, ESA SP-284, 1355–1360.

Fotheringham, A. S., Brunsdon, C. and Charlton, M. E. (2002) *Geographically Weighted Regression: the Analysis of Spatially Varying Relationships.* Wiley: Chichester, UK.

Geoghegan, J., Wainger, L. A. and Bockstael, N. E. (1997) Spatial landscape indices in a hedonic framework: an ecological economics analysis using GIS. *Ecological Economics* **23**, 251–264.

Goovaerts, P. (1999) *Geostatistics for Natural Resources Evaluation.* Oxford University Press: New York, NY, USA.

Haack, B. N., Guptill, S. C., Holz, R. K., Jampoler, S. M., Jensen, J. R. and Welch, R. A. (1997) Urban analysis and planning. In Philipson *et al.* (eds), *Manual of Photographic Interpretation*, 2nd edn. American Society of Photogrammetry: Washington, DC, USA; 517–554.

Haralick, R. M., Shanmugam, K., and Dinstein, I. (1973) Texture features for image classification. *IEEE Transactions on Systems, Man, and Cybernetics* **3**, 610–622.

Herold, M., Mueller, A., Guenter, S. and Scepan, J. (2002) Object-oriented mapping and analysis of urban land use/cover using IKONOS data. Proceedings of the 22nd EARSEL Symposium, Prague, Czech Republic, 4–6 June 2002, 531–538.

Herold, M., Gardner, M. and Roberts, D. A. (2003a). Spectral resolution requirements for mapping urban areas. *IEEE Transactions on Geoscience and Remote Sensing* **41**, 1907–1919.

Herold, M., Liu X. and Clarke, K. C. (2003b). Spatial metrics and image texture for mapping urban land use. *Photogrammetric Engineering and Remote Sensing* **69**, 991–1001.

Herold, M., Goldstein, N. C., and Clarke, K. C. (2003c) The spatiotemporal form of urban growth: measurement, analysis and modelling. *Remote Sensing of Environment* **86**, 286–302.

Herold, M., Couclelis, H., and Clarke, K. C. (2005) The role of spatial metrics in the analysis and modelling of land use change. *Computers, Environment and Urban Systems* **29**, 369–339.

Herzfeld, U. C. (1999) Geostatistical interpolation and classification of remote sensing data from ice surfaces. *International Journal of Remote Sensing* **20**, 307–327.

Jensen, J. R. and Cowen, D. C. (1999) Remote sensing of urban/suburban infrastructure and socio-economic attributes. *Photogrammetric Engineering and Remote Sensing* **65**, 611–622.

Jupp, D. L. B., Strahler A. H. *et al.* (1988) Autocorrelation and regularization in digital images. I. Basic theory. *IEEE Transaction on Geoscience and Remote Sensing* **26**, 463–473.

Kyriakidis, P. C., Liu, X. and Goodchild, M. F. (2004) Geostatistical mapping of thematic classification uncertainty. In Lunetta, R. S. and Lyon, J. G. (eds), *Remote Sensing and GIS Accuracy Assessment*. CRC Press, Boca Raton, Florida, USA; 145–162.

Lark, R. M. (1996) Geostatistical description of texture on an aerial photograph for discriminating classes of land cover. *International Journal of Remote Sensing* **17**, 2115–2133.

Liu, X. and Clarke, K. C. (2002) Estimation of residential population using high resolution satellite imagery. Proceedings of the 3rd Symposium on Remote Sensing of Urban Areas, June 2002, Istanbul, Turkey, 153–160.

Liu, X., Clarke, K. C. and Herold, M. (2005) Population density and image texture: a comparison study. *Photogrammetric Engineering and Remote Sensing* **72**(2), 187–196.

Lo, C. P. (1995) Automated population and dwelling unit estimation from high-resolution satellite images: a GIS approach. *International Journal of Remote Sensing* **16**, 17–34.

Longley, P. A. (2002) Geographical information systems: will developments in urban remote sensing and GIS lead to 'better' urban geography? *Progress in Human Geography* **26**, 231–239.

Longley, P. A. and C. Tobon (2004) Spatial dependence and heterogeneity in patterns of hardship: an intra-urban analysis. *Annals of the Association of the American Geographers* **94**, 503–519.

McGarigal, K., Cushman, S. A., Neel, M. C. and Ene., E. (2002) FRAGSTATS: Spatial Pattern Analysis Program for Categorical Maps: www.umass.edu/landeco/research/fragstats/fragstats.html [accessed 30 Jan 2005].

Mesev, V. (1998) The use of census data for urban image classification. *Photogrammetric Engineering and Remote Sensing* **64**, 431–438.

Parker, D. C., Evans, T. P. and Meretsky, V. (2001) Measuring emergent properties of agent-based landuse /landcover models using spatial metrics. Seventh Annual Conference of the International Society for Computational Economics: http://php.indiana.edu/~dawparke/parker.pdf [accessed September 2003].

Peplies, R. W. (1974) Regional analysis and remote sensing: a methodological approach. In Estes J. (ed.), *Remote Sensing: Techniques for Environmental Analysis*. Hamilton: Santa Barbara, CA, USA; 277–291.

Ridd, M. K. (1995) Exploring a V–I–S (vegetation–impervious surface–soil) model for urban ecosystems analysis through remote sensing: comparative anatomy for cities. *International Journal of Remote Sensing* **16**, 2165–2185.

Wang, F. (1990) Fuzzy classification of remote sensing images. *IEEE Transactions on Geoscience and Remote Sensing* **28**, 194–201.

Webster, R. (1973) Automatic soil-boundary location from transect data. *Mathematical Geology* **5**: 27–37.

Woodcock, C. and V. J. Harward (1992) Nested-hierarchical scene models and image segmentation. *International Journal of Remote Sensing* **13**, 3167–3187.

Zhu, C. and X. Yang (1998) Study of remote sensing image texture analysis and classification using wavelet. *International Journal of Remote Sensing* **13**, 3167–3187.

6

Using remote sensing and GIS integration to identify spatial characteristics of sprawl at the building-unit level

John Hasse

Department of Geography and Anthropology, Rowan University, Glassboro, NJ, USA

6.1   Introduction

One of the most remarkable human activities in terms of transforming and impacting the natural environment is the development of land for settlement. Patterns and configurations of urbanization have implications for a wide gamut of issues and policies, from environmental quality to health, to transportation and energy, to social and economic welfare. Global trends of rural to urban population migrations, coupled with the unprecedented technological capability of modern societies to construct urban environments, have led to magnitudes of urbanization unparalleled at any former period in history. In the USA alone, 2.08 million acres of open land were urbanized annually between 1992 and 2002 (3.95 acres/minute), an increase from 1.37 million acres/year of urbanization between 1982 and 1992 (Natural Resources Conservation Service, 2004). Not only are the rates of urban growth accelerating, but the patterns of urban growth are becoming more dispersed. The importance of urban sprawl to many public-interest, government and academic agencies has led to multiple initiatives of research and analysis. Many researchers,

Integration of GIS and Remote Sensing   Edited by Victor Mesev
© 2007 John Wiley & Sons, Ltd.

policy makers and stakeholders have an interest in monitoring, evaluating and influencing patterns of urban growth, increasing the need for a more comprehensive understanding of the phenomenon of sprawl than currently exists. Considering the land-based and spatial nature of urbanization, geospatial scientists have a significant role to play in the discourse on sprawl. Furthermore, the geospatial technologies of *remote sensing* and *GIS* are logical tools to be widely utilized for the analysis of sprawl, or problematic spatial patterns of urban growth. While geospatial research to date has only just begun to be utilized within the urban planning and policy discourse regarding sprawl, great promise exists for advancing the study and management of sprawl through the integration of remote sensing and GIS.

Since the onset of flight in the early twentieth century, remote sensing has been utilized for the delineation, analysis and evaluation of urbanization. Techniques and platforms vary widely, from film-based low-altitude monochromatic aerial photography to digital space-based hyperspectral sensors, each with particular benefits and abilities that can aid in the analysis of sprawl. Likewise, GIS has been widely utilized for urban analysis for the past several decades, greatly advanced by the creation of GIS-based demographic data by government agencies such as the US Census Bureau. Many academic sprawl-related studies utilize the US Census TIGER GIS database for various geographic extents, such as metropolitan areas (MAs) and urbanized areas (UAs), as well as census tracts and census blocks. Because remote sensing and GIS techniques and technologies have become so closely inter-related, it is now possible to seamlessly utilize both within the same computing environment. However, this ease of integration has only recently become available. In the past, urban research has tended to develop along two largely separate tracks, one following a more demographic approach (primarily GIS-based) and the other following a more physical/environmental approach (primarily remote sensing-based). As these two tracks continue to merge and become integrated, both technologically and methodologically, new methods become available for researchers to more effectively delineate, analyse and understand the patterns and processes of sprawl.

6.2   Sprawl in the remote sensing and GIS literature

Past studies of sprawl can be divided into two general camps, *physical landscape-based analysis* and *demographic-based analysis*. Remote sensing has been most often employed in physical approaches to analysing sprawl, due to its ability to provide temporal/spatial information on the physical covering of the Earth at a given time period. The usefulness and potential application of remote sensing for urban analysis has steadily grown with the increasing numbers of remote sensing platforms, decreasing costs and ever-increasing sophistication of computer techniques. This point was recently highlighted by several prominent remote sensing journals that dedicated entire issues to focus solely on urban themes, e.g. *Remote*

Sensing of the Environment 2003; **83**(3), and *Photogrammetric Engineering and Remote Sensing* 2003; **69**(9).

Remote sensing literature has tended to use the term 'sprawl' as related to urbanization somewhat loosely, often to indicate rapid urbanization, or urbanization along the urban/rural fringe, or low-density urbanization (Hurd *et al.*, 2001; Weng, 2001; Epstein *et al.*, 2002). Classic change-detection techniques utilizing multi-date imagery have been one common approach for identifying newly developing areas of low-density urbanization (e.g. Civco *et al.*, 2002). Other remote sensing approaches have utilized night-time lights as a proxy for urban extent to iden-tify low-density sprawl (Sutton, 2003; Cova *et al.*, 2004). However, these remote sensing approaches thus far arguably lack meaningful application to the processes and patterns responsible for sprawl.

GIS-based studies of sprawl have tended to use the term more precisely than has the remote sensing literature. A number of seminal sprawl-measurement studies have occurred in recent years that utilized a primarily GIS demographic approach. Several papers have utilized population density-based metrics to provide cross-comparisons and rankings for multiple metropolitan areas within the USA (Fulton *et al.*, 2001; Nasser and Overberg, 2001; Lopez and Hynes, 2003). Many of these approaches utilize US Census Bureau data for MAs, which consists of the coun-ties with population and commuting ties to a major city. Other studies have used the US Census Bureau's UAs, which are incorporated areas and census designated places of 2500 or more persons. For example, Galster *et al.* (2001) utilized US Census metropolitan data variables for calculating their eight measures of sprawl. Theobald (2001) developed metrics for *rural* sprawl based on population densities in census tracts specifically outside of urban areas. Sprawl analytical methods employed thus far have tended to utilize either a primarily vector GIS-based or primarily remote sensing-based approach. We will come back to this point later in the chapter and unite GIS and remote sensing as we explore the most recent progress in sprawl research. However, we first must tackle one of the confounding issues in the sprawl discussion, namely, what exactly is being discussed? How do people view the idea of sprawl?

6.2.1 Definitions of sprawl

Many books have been written and studies conducted on various aspects of urba-nization. However, the term 'sprawl' is often incorrectly used as a synonym for *urban growth* in general. The identification of sprawl as a specific type and potentially problematic pattern of urbanization first arose in public discourse in the middle of the twentieth century, when suburban subdivisions began to arise in areas peripheral to existing urban locations (Hess *et al.*, 2001). To the lay person the term 'urban sprawl' is generally used to refer to spreading suburban development patterns associated with repetitive housing tracts, strip shopping malls and increased traffic congestion.

In recent decades the term has tended to be more indiscriminately used. Any development unwanted by a particular interest is often labelled as 'sprawl', regardless of the fact that it may actually embody characteristics of *smart growth* (the catch phrase for urbanization that is well-designed and non-sprawling), such as high-density, in-fill and mixed use. This inconsistent and sometimes contradictory use of the term 'sprawl' creates a risk that the word will become hackneyed or outright meaningless. In order for the phenomenon of sprawl to be adequately delineated, analysed and managed, a more precise and universally agreed-upon meaning needs to be established.

In the past several decades the interest in sprawl, and consequently the number of research articles focusing on sprawl, has risen across multiple disciplines, from public policy to environment to land management. The academic literature of urban sprawl has itself sprawled into what is characterized by Galster *et al.* (2001) as an ambiguous 'semantic wilderness'. Galster *et al.* categorize the literature into six groups of definitions that look at sprawl in the following ways: (a) sprawl defined by example; (b) sprawl defined by aesthetic definition; (c) sprawl as the cause of an unwanted externality; (d) sprawl as a consequence; (e) sprawl as selected patterns of land development; and (f) sprawl as a process of development of land use. Any use of geospatial technologies to assist in sprawl research will be more effective if it can be based on a clear definition. While sprawl may have many non-spatial socio-economic characteristics, remote sensing and GIS are spatial technologies and therefore are most useful with a definition based on the spatial pattern, extent and configurations that urbanization takes upon a landscape.

By most definitions, sprawl is a pattern of urbanization that carries with it inherent problems, dysfunctions and inefficiencies (Burchell *et al.*, 1998; Ewing, 1997; Johnson, 2001). The urban planning and policy literature provides a number of references to sprawl that help to define it in terms of a specific spatial form of urban growth. Reid Ewing (1997) offers a summary of 17 references to sprawl in the literature as being characterized by 'low-density development, strip development and/or scattered or leapfrog development'. Ewing also uses a transportation component to help define sprawl. He suggests that the lack of non-automobile access is also a major indicator of sprawl. Burchell and Shad (1999) present a working definition of sprawl as 'low-density residential and nonresidential intrusions into rural and undeveloped areas, and with less certainty as leapfrog, segregated, and land consuming in its typical form'. Consensus is emerging that sprawl is complex and cannot be characterized as a singular homogeneous phenomenon, but instead has multiple possible characteristics. Furthermore, sprawl is different from place to place (Burchell *et al.*, 1998) and can be grouped into at least three different families relating to *urban sprawl*, *suburban sprawl* and *rural/exurban sprawl* (Hasse, 2004; Theobald, 2004). Many other papers refer to sprawl as urbanization with specific spatial characteristics (Table 6.1).

The discourse on *smart growth* also helps to inform the development of sprawl measures, because the spatial characteristics of smart growth are in some respects the

Table 6.1 Spatial characteristics of sprawl found in the literature

Characteristic	Description	Selected references
High/inefficient land consumption; low population density	Low population density; high levels of urbanized land per person; rate of land urbanization greater than rate of population growth, especially in fringe areas	Black, 1996; Downs, 1998; Freeman, 2001; Galster *et al.*, 2001; Harvey and Clark, 1965; STPP, 2000; Montaigne, 2000; Hasse, 2003
Fringe development	Development away from city centre; rapid development of open spaces on city boundary	Besl, 2000; Downs, 1998; Galster *et al.*, 2001; Katz and Bradley, 1999
Lack of connectivity	Arterial street systems; lack of grid; lots of dead ends	Duany and Plater-Zyberk, 1998; NRDC, 1996; Hasse, 2003
Leapfrogging; scattered development	Development that skips over empty parcels	Clawson, 1962; Mills, 1981; Downs, 1998; Gordon and Richardson, 1997b; Yeh and Li, 2001; Hasse, 2003
Separation of uses	Different land uses (employment, retail, residential) are far apart; residential development beyond edge of employment and retail services; lack of residential development in city centre	Brown *et al.*, 1998; Downs, 1998; Duany and Plater-Zyberk, 1998; Ewing, 1994, 1997; Galster *et al.*, 2001; Hasse, 2003
Lack of functional open space	Lack of open space that performs a useful public function; ill-defined residual space	Anonymous, 1999; Ewing, 1997, 1994; Hasse, 2003
Lack of non-auto transportation accessibility	Dispersed spatial patterns and long distances to destinations preclude use of public transit, bicycle and pedestrian modes of travel.	Downs, 1998; Ewing, 1997, 1994; Hasse, 2003
Aesthetics and architecture	You know it when you see it. Big-box retail; strip malls; no sidewalks; excessively wide roads. Large, disjointed buildings set back from street, highly articulated, rotated on lots	Duany and Plater-Zyberk, 1998; Gore, 1998; Koffman, 1999; Kunstler, 1996; NRDC, 1996; Hasse, 2003

Adapted and modified from Hess *et al.* (2001).

mirror opposites of the characteristics of sprawl. According to the US Department of Environmental Protection, smart growth principles promote development which:

> ... has mixed land uses; takes advantage of compact building design; creates a range of housing opportunities and choices; creates walkable neighborhoods; fosters distinctive, attractive communities with a strong sense of place; preserves open space, farmland, natural beauty, and critical environmental areas; strengthens and directs development towards existing communities; provides a variety of transportation choices; makes development decisions predictable, fair, and cost effective; and encourages community and stakeholder collaboration in development decisions. (US EPA, 2005)

The spatial patterns of smart growth and sprawl are inherently different and able to be distinguished at various scales through appropriate geospatial methods.

6.2.2   Spatial characteristics of sprawl at a metropolitan level

A number of spatial-based measurements designed to capture various sprawl signatures have evolved out of the characteristics of sprawl listed in Table 6.1. Torrens and Alberti (2000) explored developing an empirical landscape framework to sprawl measurement that focuses on the characteristics of *density*, *scatter*, the *built environment* and *accessibility*. They outlined a set of metrics for quantifying these characteristics that employ *density gradients*, *surface-based approaches*, *geometrical techniques*, *fractal dimensions*, *architectural and photogrammetric techniques*, *measurements of landscape composition and spatial configuration*, and *accessibility calculations*. One of the seminal works of spatial measurements of sprawl at the metropolitan level was developed by Galster *et al.* (2000), who define sprawl as 'a pattern of land use in an urbanized area that exhibits low levels of some combination of eight distinct dimensions: density, continuity, concentration, compactness, centrality, nuclearity, diversity, and proximity' (Galster *et al.*, 2001). They operationalized six of these indicators to compare the characteristics of sprawl for 13 metropolitan areas in the USA. Figure 6.1 portrays the schematic diagrams from Galster *et al.* (2001), demonstrating the spatial patterns captured by each metric for sprawling and non-sprawling metropolitan areas.

A number of other studies have also taken a GIS-based approach to develop sprawl measures for comparing metropolitan areas. Malpezzi (1999) analysed the spatial distribution of population within census tracts of US Metropolitan Statistical Areas (MSAs), calculating various indices of *density* as well as *commuting patterns*. Ewing, Pendall and Chen (2002) developed an index for sprawl which combined individual measures for: *residential density*; *neighbourhood mix of homes, jobs and services*; *strength of activity centres and downtowns*; and *accessibility of the street network*. Hess *et al.* (2001) developed a suite of seven spatial metrics for sprawl that focused on *land consumption, population concentration, separation of land*

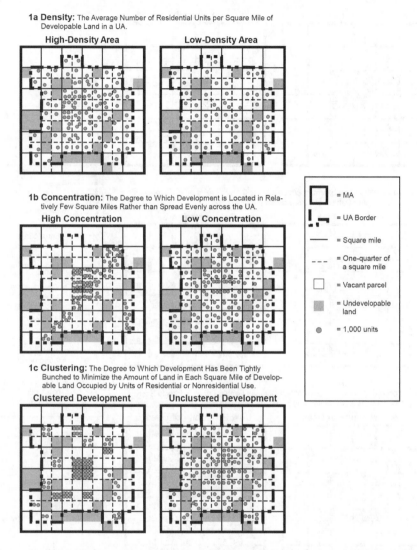

Figure 6.1 Metropolitan-level spatial measure of sprawl. Galster *et al.* (2001) utilized US Census metropolitan areas (MAs) and urbanized areas (UAs) data to operationalize six measures of sprawl at the metropolitan level, including: (a) density; (b) concentration; (c) clustering; (d) centrality; (e) nuclearity; and (f) proximity. Reproduced from Galster *et al.* (2001) Wrestling sprawl to the ground: defining and measuring an elusive concept. *Housing Policy Debate* **12**, 681–717, courtesy of the Fannie Mae Foundation

1d Centrality: The Degree to Which Development in a UA is Located Close to the CBD.

Highly Centralized Area   **Highly Decentralized Area**

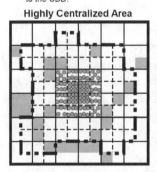 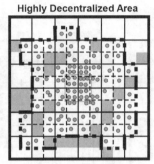

1e Nuclearity: The Extent to Which a UA is Characterized by a Mononuclear or Polynuclear Pattern of Development

Monoclear Area   **Polynuclear Area**

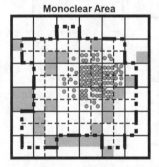 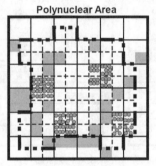

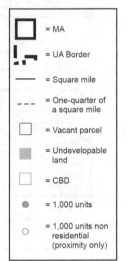

1f Proximity: The Degree to Which Different Land Uses are Close to Each Other Across a UA. (note: grey circles denote residential only)

High Proximity of Uses   **Low Proximity of Uses**

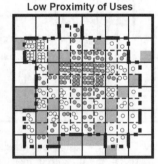

Figure 6.1   (Continued)

uses/accessibility, and *temporal patterns of sprawl*. They calculated their metrics for 49 urbanized areas within the USA, finding little correlation between the measures, suggesting that sprawl has a heterogeneous spatial nature on an interurban scale.

6.2.3   Spatial characteristics of sprawl at a submetropolitan level

The studies covered thus far have been conducted on a metropolitan scale, providing a single value index to characterize certain aspects of sprawl for an entire urban region. A comparison of the results for various cities is interesting and sometimes surprising (alas, Los Angeles is not even close to being the most sprawling city in the USA). However, some researchers question how much meaning to place on these measures, as well as how valuable such measures are to inform policy decisions (Hess *et al.*, 2001; Hasse and Lathrop, 2003b; Song and Knaap, 2004). As argued by Hasse and Lathrop (2003b), there is likely much more variation in sprawling urbanization within any particular metropolitan area than exists between different metropolitan areas. Some of the most recent sprawl analysis work has focused on submetropolitan measures of sprawl. Song and Knaap (2004) derived a set of neighbourhood-scale sprawl measures adapted from a planning support software system called INDEX, developed by Allen *et al*. Song and Knaap operationalized five measures of urban form, including: *street design and circulation systems*; *density*; *land use mix*; *accessibility*; and *pedestrian access* for 186 neighbourhoods in metro-Portland, Oregon. Utilizing census blocks as a proxy for neighbourhoods, Song and Knaap focused on two neighbourhoods, one that embodied the characteristics of new urbanism (the so-called 'smart growth') and the other that represented Portland's average suburban tract. Song and Knaap also conducted a correlation analysis of their measures, by the median age of neighbourhood housing stock, to establish the change in sprawling characteristics of Portland over time.

At the submetropolitan level, the problematic characteristics of sprawl can be more systematically identified and measured than at the metropolitan level. Hasse (2004) created a set of 12 geospatial indices of urban sprawl (GIUS), designed specifically to provide information about what characteristics are considered problematic or dysfunctional for an individual development (Table 6.2). The GIUS measurements were utilized to evaluate and compare three recently constructed housing tracts within a county on the rural/urban fringe of New Jersey. The GIUS metrics are micro-measures of sprawl that provide quantitative information for individual development tracts for three categories of characteristics: (a) *land-use patterns*; (b) *transportation patterns*; and (c) *environmental impact patterns*. The GIUS metrics employ various GIS-based spatial measurements of landscape parameters identifiable in land use, road networks and various environmental mapping sources. Six of the GIUS measures are provided in schematic form for two scenarios of a fictitious town; one scenario with sprawl and the second scenario with smart growth (Figure 6.2).

Table 6.2   Twelve tract-level GIUS measure of sprawl

Measure	Description	Calculation
1. Density	Measures the intensity of land utilization for a given tract	Areal size of tract divided by number of housing units within tract
2. Leap-frog (Figure 6.2a)	Measures the degree to which new tracts skip over vacant parcels adjacent to previous settlement	Straight line distance from new tract to previous settlement
3. Segregated land use	Measures the degree to which new tracts are mixed with other categories of urban land use	Count the number of different categories of urban land use within a 1500 ft buffer (i.e. 10 minute walk) to new tract
4. Regional planning inconsistency (Figure 6.2b)	Indicates whether a new tract is inconsistent with regional and state plans	Tract is assigned a weighted value dependent on its location within a regional plan
5. Highway strip (Figure 6.2c)	Indicates whether a new tract is situated in strips fronting along rural highways	Tract is overlaid with a 500 ft buffer of rural highways
6. Road infrastructure inefficiency	Measures the inefficiency of road infrastructure by measuring road length, number of intersections and cul-de-sacs of new development tracts	Length of road, number of intersections and number of cul-de-sacs are summed by tract and divided by the number of units within the tract
7. Transit inaccessibility	Measures the degree to which non-auto modes of travel are accessible to new tracts	Calculates road distance from tract to pedestrian/bicycle routes and public transportation stops
8. Community node inaccessibility (Figure 6.2d)	Measures how scattered a new tract is from important community centres such as schools, libraries, fire/rescue, police, recreational facilities, etc.	Calculates road distance from tract to a set of nearest community nodes
9. Consumption of important land resources (Figure 6.2e)	Measures the degree to which new tracts consume important agricultural and natural land resources	Calculates the area of prime farmland, core forest habitat and wetlands displaced by tract and divides by the number of units

10. Sensitive open space encroachment	Measures the proximity of new tract to sensitive open space, including documented threatened/endangered wildlife habitat and preserved farmland	Calculates the distance of tract to nearest wildlife habitat and preserved farm parcels
11. Impervious surface coverage (Figure 6.2f)	Measures the amount of impervious surface imposed from a given tract	Calculates the total area of impervious coverage of a tract and divides by the number of units within the tract
12. Growth trajectory	Measures the pace of growth in terms of new development and locality size and remaining available land	Calculates the percentage of urban spatial increase in terms of: (a) previous urban extent; (b) municipal size; (c) remaining available land

Adapted from Hasse (2002).

The GIUS measures were operationalized for Hunterdon County, New Jersey, for all housing tracts constructed county-wide between 1986 and 1995 (Hasse, 2004). To demonstrate the functionality of the GIUS measures, three development tracts were selected that epitomized the most sprawling, average and smartest-growing development that occurred, as measured by the GIUS metric (Figures 6.3a–c). The study established that many of the spatial characteristics of sprawl can be meaningfully quantified and compared at the micro-level of individual housing tracts (Figure 6.4).

6.3   Integrating remote sensing and GIS for sprawl research

While Hasse's GIUS sprawl indices (2004) are primarily spatial-based measurements and therefore might be placed within the GIS- based camp of sprawl analysis, many of the data utilized by Hasse were originally derived from remote sensing-based data sources, such as digital orthophotography, making this work a substantial integration of remote sensing and GIS. Many of the GIUS measures could be adapted to other platforms of remote sensing- and raster-based analysis.

A number of other recent works in sprawl research rely more substantially on combining both GIS and remote sensing technologies and techniques. Analytical approaches that integrate remote sensing and GIS technologies are able to provide a more robust and sophisticated line of attack than either technology can provide in isolation. Software advances are facilitating the ease with which researchers are able to integrate vector-based GIS, raster-based GIS and remote sensing techniques. There are substantial benefits to integrating the physical land use/land cover information provided by remotely sensed data and the growing body of socio-economic and infrastructure information available for GIS.

SMART GROWTH PATTERN   **SPRAWL PATTERN**

2a Leapfrog: The Straight Line Distance from a Previous Settlement to a
New Development Tract. Longer Distances are More Sprawling.

 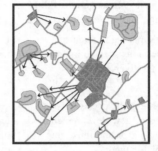

2b Regional Planning Inconsistency: The Location of a Tract within a
Regional Plan. Tracts not within Planned Growth Areas are Sprawl.

 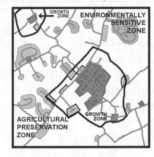

▣	= Previously Existing Settlement
▢	= New Tract (Patch) of Development
╱	= Road
↗	= Leapfrog Distance
╱	= Growth Boundary
·-·-	= Rural Highway Buffer
▮	= Highway Strip Tract (Patch)

2c Highway Strip: The Location of a Development Tract within a Rural
Highway Buffer. Tracts within the Rural Highway Buffer are Sprawl.

 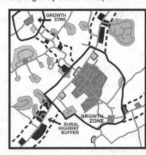

Figure 6.2 Development tract-level spatial measures of sprawl. Hasse (2004) developed 12 geospatial indices of urban sprawl (GIUS) at the development tract level. These conceptual schematic diagrams illustrate selected GIUS measurement for a fictitious town that grows with a smart growth pattern (left) and sprawl pattern (right). The measurements selected include: (a) leapfrog; (b) regional planning inconsistency; (c) highway strip; (d) community node inaccessibility; (e) land resource impacts; and (f) impervious surface coverage. Reproduced with permission of the University of Wisconsin Press from Hasse, J.E. (2004) A geospatial approach to measuring new development tracts for characteristics of sprawl. *Landscape Journal: Design, Planning and Management of the Land* **23**, 1–4

SMART GROWTH PATTERN   **SPRAWL PATTERN**

2d Community Node Inaccessibility: The Road Distance from a Tract to a Nearest Set of Selected Community Destinations (schools, churches, grocery, etc). Larger Average Distances are Sprawling.

2e Land Resource Impacts: The Amount of Prime Farmland, Habitat and Wetlands Consumed by a Tract. Tracts with Greater Per Capita Resource Consumption are Sprawling.

■ = Previously Existing
     Settlement

▨ = New Tract (Patch)
     of Development

╱ = Road

╌╌▸ = Road Distance to
     Community Nodes

o = Community Node
     (school, church etc.)

▧ = Prime Farmland

⬡ = Wildlife Habitat

⬡ = Wetlands

▨ = Land Resource
     Impact of New Tract

■ = New Tract
     Impervious Surface

2f Impervious Surface Coverage: The Total Amount of Impervious Cover Created by a Tract of Development. Sprawl Creates Larger Amounts of Impervious Surface Per Resident or Per Employee.

Figure 6.2   (Continued)

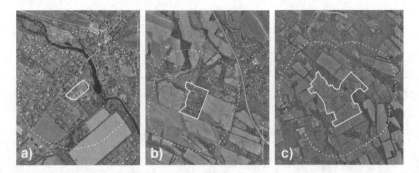

Figure 6.3 Selected development tracts for demonstrating GIUS. These three tracts of suburban development were selected from a countywide GIUS analysis of new development. The tracts have been named for the municipality in which they were located: (a) Califon; (b) Readington; and (c) Alexandria. Each tract is delineated by a solid white line and a dashed 1500 ft pedestrian accessibility buffer. Reproduced with permission from Hasse, J. E. (2004) A geospatial approach to measuring new development tracts for characteristics of sprawl. *Landscape Journal: Design, Planning and Management of the Land* **23**, 1–4

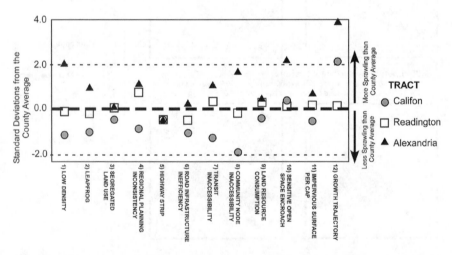

Figure 6.4 Normalized GIUS measures for three selected tracts. This graph depicts the value of each GIUS metric in standard deviations from the county average. While the three selected tracts effectively demonstrate lower than average, average and higher than average sprawl values in the county for most of the variables, the measure are not highly redundant. Many other development tracts within the county had a broad mixture of values. Reproduced from Hasse, J. E. (2002) Geospatial Indices of Urban Sprawl in New Jersey. Doctoral Dissertation, Rutgers University, New Brunswick, NJ, USA; 224 pp.

The most basic category of GIS integration with remote sensing is land use mapping derived from remotely sensed sources. For example, a number of sprawl-related studies conducted in New Jersey (Hasse and Lathrop, 2001, 2003a; MacDonald and Rudel, 2004) utilize the state's highly detailed digital land use/land cover database, which was delineated statewide from on-screen digitizing of digital orthophotography (Thornton *et al.*, 2001). While the analysis relied heavily on vector-based GIS techniques to measure temporal landscape changes, the data layers required for the calculations included *land use/land cover, impervious surface, fresh water wetlands*, and *prime farm soils*. Each of these data layers used remotely sensed imagery as its primary source.

Some approaches to sprawl research have utilized a primarily remote sensing approach augmented by various ancillary GIS data or GIS spatial methodology. For example, Yeh and Li (1998, 2001) used remotely sensed data to measure and monitor the degree of urban sprawl for cities and towns in China, using an entropy measure of dispersal along roads. Sudhira *et al.* (2004) integrated IRS 1C and LISS multispectral imagery with Survey of India (SOI) topo-sheets to develop temporal metrics of sprawl in Karnataka, India. While these studies are somewhat ambiguous in making a clear distinction between specific characteristics of sprawl and urban growth in general, they demonstrate the utility of augmenting large-scale remote sensing platforms with ancillary GIS data, such as overlaying vector-based roads with digital imagery to better evaluate urban processes related to sprawl.

A more sophisticated analysis of sprawl, utilizing the European CORINE land cover dataset, which was compiled from multiple satellite imagery and ancillary GIS sources, was conducted for 15 cities within Europe (Kasanko *et al.*, 2005). Five indicator sets were developed to shed light on whether European cities were experiencing a dispersion of population density, by examining *residential land use, land taken by urban expansion, population density* and *urban density*. The team found that European cities were becoming more dispersed in general but that there were also significant differences in the densities of growth between southern, eastern and north-western cities.

One of the problematic characteristics of sprawl is the wasteful consumption of important natural resources. Sprawling development patterns impose a large ecological footprint by moving a relatively small number of residences into large-lot housing. The integration of remote sensing and GIS can facilitate the study of natural resource impacts attributable to sprawl. Hasse and Lathrop (2003a) developed a set of 'land resource impact' (LRI) indicators that measured the per capita population impact of sprawling urbanization on five specific critical land resources, including: (a) *urban density* (i.e. efficiency of land utilization); (b) *prime farmland loss*; (c) *core forest habitat loss*; (d) *natural wetlands loss*; and (e) *impervious surface cover gain*. By integrating demographic census data with landscape change data, the authors were able to demonstrate impacts on a per-capita basis, in order to illustrate that sprawling development patterns consume more resources for each person provided with housing than do smart growth patterns. The five measures were calculated

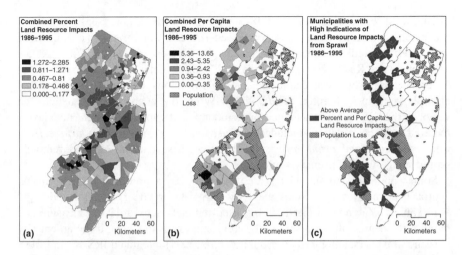

Figure 6.5 Land resource impact indicators of sprawl in New Jersey. Sprawl consumes significant quantities of important land resources including: prime farmland, forest core habitat, and freshwater wetlands. These maps depict the municipalities that: (a) lost the greatest percentage of these resources; (b) lost the greatest amounts of the resource per person added to the population; and (c) have both high percentage and per capita loss. Reproduced with permission from Hasse, J. E. and Lathrop R. E. (2003b) Land resource impact indicators of under sprawl. *Applied Geography* **23**, 170. ©Elsevier (2003)

on an individual municipal basis and then combined into an index that provides an overall indication of the municipalities in which sprawl is having the greatest impact on critical land resources (Figure 6.5). The data utilized for this analysis were derived from remotely-sensed sources, such as orthophotography for the land use/land cover and wetlands delineation (Thorton *et al.*, 2001). The prime farm-soils soil maps were generated by the US Natural Resources Conservation Service on a county basis, and originally derived from aerial photography, geological maps and in-field samples. Lathrop (2004) updated the statewide analysis by incorporating new development polygons screen-digitized from SPOT imagery.

The approach to sprawl that focuses on the physical environment also includes a substantial literature of ecology-based studies that often employ remote sensing techniques to characterize the degree of urban intensity within a landscape ecology context (Jensen *et al.*, 2004; Forys and Allen, 2005; MacDonald and Rudel, 2005; Theobald, 2004). The FRAGSTATS software package (McGarigal and Marks, 1995), widely used to generate landscape-based metrics for landscape ecology (Gustafson, 1998), is now being applied to urban analysis. Herold *et al.* (2005) explored a framework for combining remote sensing with these landscape ecology metrics in order to improve the analysis and modelling of urban growth and land use change. The authors demonstrated through a pilot study of the Santa Barbara, California, coastal area that the combination of remote sensing GIS-based spatial metrics can contribute an important new level of information to urban modelling and urban dynamic analysis. This line of landscape-scale (i.e. tract-level or patch-level) GIS-remote sensing integration for urban analysis holds great potential for moving beyond some of the past limitations of modelling urban dynamic process and specifically urban sprawl.

Meaningful integration of remote sensing data with spatial metrics for measuring sprawl is also beginning to occur in some of the urban planning and geography literature. The previously discussed work of Galster *et al.* (2001; Figure 6.1) broke new ground in developing sprawl spatial measurements by converting census-based GIS data into a grid. The Galster study developed a number of spatial metrics with some similarities to landscape ecology metrics by creating half-mile and 1-mile grids of the census data polygons. Wolman *et al.* (2005) argued that the methodology of Galster *et al.* (2001) was limited in several respects, including its inability to compensate for land that was impossible to develop when calculating various density measurements. Wolman improved on Galster *et al.*'s methods by integrating land use data from the US Geological Survey's (USGS) National Land Cover Database (NLCDB). The NLCDB is a nationwide land-use map derived from remotely sensed satellite imagery at 30 m resolution. Wolman's integration of land cover data demonstrably changed Galster *et al.*'s density measures from as little as 2.6 to as much as 27.1 for selected metropolitan areas, although very little change in rank occurred from Galster *et al.*'s original study. The integration of remote sensing for updating land use/land cover information in sprawl analysis will continue to mature as sprawl metrics are refined and the ease with which timely ground data can be added to the analysis improves.

One of the problems interfering with a more substantial use of geospatial technologies (especially remote sensing) within urban research is that many of the metrics and analyses thus far developed have had a poor relationship to urban spatial theory and/or application in policy making. The development of sprawl measurements that can take advantage of the benefits of integrating remote sensing and GIS needs to be applicable to planners in the trenches. One of the places in which there is great potential for geospatial science, landscape metrics and planning and policy to mutually enhance one another is the topic of sprawl. Developing better digital representations of the urban process requires exploration of the urban process at its most fundamental scale.

6.4 Spatial characteristics of sprawl at a building-unit level

One area of research that holds promise for advancing urban analysis and urban sprawl also opens new avenues for integrating remote sensing with GIS. By breaking down urban processes to the most fundamental units, the basic building blocks of urban organization can be reproduced within a digital environment. 'Urban atomization' entails rethinking how to represent and model the urban phenomenon within a GIS at the most fundamental urban unit. Typically, urban social analysis has tended to occur within a vector GIS digital environment, while environmental/landscape analysis has tended to utilize raster-based approaches. While each method has its advantages and disadvantages for modelling landscape structure, there are nevertheless still many limitations with both raster and vector analytical approaches related to issues of scale, temporal change, data conversion and ecological fallacy/modifiable areal unit problem (MAUP) Openshaw 1984a, 1984b) among many others. It can be awkward at best to represent many aspects of urban processes in either a solely-raster or solely-vector data platform. In order to move beyond these limitations, it may be advantageous to represent urban phenomena by reducing urban structure down to the smallest basic elements.

Instead of trying to fit the urban process into raster cells or polygons, researchers are asking how to best model the fundamental components of the urban process within state-of-the-art geospatial digital environments. Considering that the urbanization process consists of the nexus between the physical built environment and social processes, a robust GIS urban modelling environment should be built upon the most basic fundamental unit or smallest elements by which the urbanization process functions. Demographic data are often available to researchers at the metropolitan, neighbourhood, census block and zip code level, making these spatial units logical choices for analysis of sprawl thus far highlighted throughout this chapter. In contrast, the social units by which demographic data are collected through surveys and censuses are often the individual person living within the city, the family and the household, but these data are protected from public disclosure due to issues of privacy. The urban process is complex and dynamic and consists of a combination of the physical urban structure and the social structure of the people living in and using the city. Since individuals, families and households are highly transitory, it can be argued that *building units* emerge as the logical fundamental or smallest solid 'atom' of urban spatial structure.

By modelling urban spatial structure as elemental building units that exist at a particular time and location in space, building units become the 'urban atoms' of a data structure that can then be organized and combined into a nested hierarchy of functional entities at the appropriate scale for the phenomenon of interest. To use a biological analogy, building units can be viewed as the most basic *cells* of urban structure. Neighbourhoods can be conceptualized as logical groupings of building unit cells into discrete functional areas or the 'organs' of the urban organism. Neighbourhoods linked together through transportation and infrastructure networks become the functional urban systems. The city itself combines the various neighbourhoods and systems into the complete functioning (or sometimes dysfunctioning) urban organism.

New GIS data structures, such as the ESRI Geodatabase, hold potential for innovative nested hierarchal approaches to urban geospatial data modelling. Individual components of the atomic urban data model can be modular and object-orientated, so that each building unit can 'know' its own location, statistical summaries of the people living/employed in the building, the land area occupied and the building floor area, available social and health-related data, etc. Object-orientated building units could also contain information about their own date of creation and thus be incorporated into temporal modelling of urbanization. Urban data structure could become hierarchical, meaning that, depending on the scale of interests, building units could be represented as points, polygons or triangular irregular networks (TINs), and multiple units could be grouped into regions to represent a neighbourhood or interpolated into a surface to visualize particular variables, etc. Atomic urban data structure will also facilitate new approaches to integrating remote sensing data with object-orientated GIS data, substantially advancing all branches of urban analysis, including sprawl.

Work is just beginning on an urban atomization approach that integrates remote sensing with building unit locations. Mesev (2005) is exploring the use of postal points, which are GPS building location points generated by the Ordnance Survey of Great Britain that map the building centroid of commercial or residential buildings with postal delivery. This dataset is updated four times a year and provides a highly accurate spatial inventory of building units. Mesev integrates these postal points with IKONOS imagery to examine spatial patterns of residential neighbourhoods and commercial areas. Groups of these points were used to characterize the spacing and arrangement of residential and commercial buildings, using nearest-neighbour and linear nearest-neighbour indices. Although the pilot analysis explored only two UK cities for two relatively non-complex variables, including *density* (compactness vs. sparseness) and *linearity*, Mesev argues that multiple avenues of research can emerge, such as automated pattern recognition through building unit integration with remote sensing imagery.

6.5 A practical building-unit level model for analysing sprawl

Hasse and Lathrop (2003b) utilized an urban atomization approach to evaluate several characteristics of sprawl by measuring sprawl characteristics for individual housing units. Hasse and Lathrop contended that a housing-unit approach to measuring sprawl is the most meaningful because each house can have a different performance of sprawl and smart growth. By generating measures at the atomic (housing-unit) level, Hasse and Lathrop were able to rescale the data up to any geography of interest, such as a housing tract, census block or municipality. This effectively solved a number of rescaling and overlay issues and limitations. Hasse and Lathrop's method for locating each housing unit was accomplished by intersecting remote sensing-derived urban land use/land cover classified regions with digital parcel maps and generating centroids for the resulting polygons

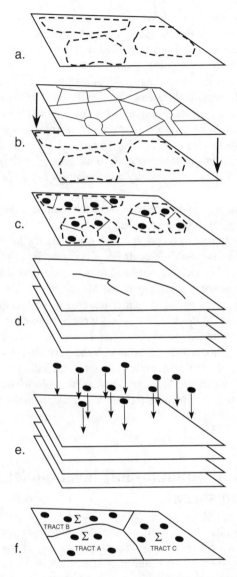

Figure 6.6   Delineation of housing unit locations through the integration of GIS and remote sensing. Household locations are delineated as vector point locations through a multi-step process: (a) delineation of new urbanization (image classification or heads-up digitizing); (b) intersection of new development patches with digital parcel map; (c) polygon centroids estimate location of new housing unit; (d) generation of various sprawl parameters, e.g. density, leapfrog, segregated land use, highway strip, and community node inaccessibility; (e) assignment of various sprawl parameters to housing unit point theme; (f) summary of individual housing unit metric values by regions of interest, such as census tracts or municipalities

(Figure 6.6). This technique is particularly necessary in rural areas, where housing unit locations are unlikely to be aligned with the tax parcel's physical centroid. The resulting point dataset is an accurate estimate of each housing unit location (Figure 6.7).

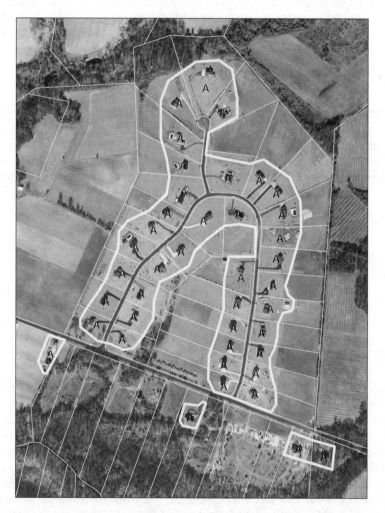

Figure 6.7 Housing unit location automation. This image depicts an orthophoto of one newly developed housing tract. The thick lines delineate the 'patches' of new urban growth as classified by the land use/land cover dataset. The thin lies delineate the property parcel lines. The target symbol denotes the automated centroid location estimated for each new housing unit. Sprawl measurements are calculated for each housing unit centroid

Although most of the 12 GIUS measures developed on a tract-level can be applied to the housing-unit scale, five measures are described here in detail, including: *density, leapfrog, segregated land use, community node inaccessibility* and *highway strip,* The calculations are made using various GIS techniques and the corresponding values are assigned to each new housing unit for the set of five selected metrics. The data are then scaled-up to municipality by summarizing the housing points within each municipal boundary, in order to provide a 'sprawl report card' for recent growth for each locality. The following section details the Hasse and Lathrop housing unit level methodology (from Hasse and Lathrop, 2003b).

6.5.1 Urban density

The urban density indicator provides a measure of the amount of land area occupied by each housing unit (Figure 6.8a). The municipal urban density (UD_{mun}) was calculated by summing the land areas for each new housing unit and dividing that sum by the total number of units within each municipality, as depicted in equation 6.1. Lower density indicates a sprawling signature for the density measure.

$$UD_{mun} = \frac{\sum DA_{unit}}{\sum N_{unit}} \tag{6.1}$$

where:
UD_{mun} = urban density index for new urban growth within a municipality,
DA_{unit} = developed area of each unit, and N_{unit} = number of new residential units.

6.5.2 Leapfrog

Tracts of urban growth that occur at a significant distance from previously existing settlements are considered 'leapfrog' (Figure 6.8b). The leapfrog indicator was calculated by measuring the distance from the location of each new housing unit (at time 2) to previously settled areas (at time 1). The previous settlements were delineated as tracts of urban land use existing in time 1 that corresponded to designated place names on USGS quadrangle maps or existing tracts larger than 50 acres (20.23 hectares). This process filtered out smaller non-named tracts of time 1 urban areas that had already leapfrogged from settled areas. A straight-line distance grid was generated from these 'previously settled' tracts and the grid value was assigned to each new housing unit. The housing-unit leapfrog value was then scaled to the municipal leapfrog index (LF_{mun}) by summarizing the leapfrog field value of the housing-unit point layer by municipality, as depicted in equation 6.2. New growth that occurs at large leapfrog distances is considered

SMART GROWTH   **SPRAWL**

8a Density: Defined as the Amount of Land Consumed Per Building Unit.

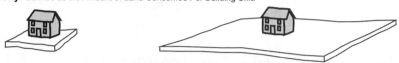

8b Leapfrog: Defined as the Distance between a New Building Unit and the Nearest Edge of an Previously Existing Settlement.

8c Segregated Land Use: Defined as the Number of Different Land Uses within a 10 Minute Walking Distance (1,500 ft) to a Building Unit.

8d Highway Strip: Defined as the Location of a Building Unit Within a Buffer Along Rural Highways.

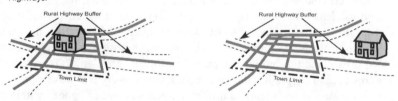

8e Community Node Inaccessibility: Defined as the Average Road Distance from a Building Unit to the Nearest Set of Selected Communities Destinations.

Figure 6.8 Conceptual diagrams for housing unit sprawl measures. Sprawl measurements are conducted for individual housing units for selected characteristics, including: (a) density; (b) leapfrog; (c) segregated land use; (d) highway strip; and (e) community node inaccessibility. Other sprawl characteristics are also measurable at the housing-unit level, which facilitates scaling to any geography of interest

sprawling.

$$LF_{mun} = \frac{\sum Dlf_{unit}}{\sum N_{unit}}$$ (6.2)

where LF_{mun} = leapfrog index for new urban tracts within a municipality, Dlf_{unit} = leapfrog distance for each new unit, and N_{unit} = number of new residential units.

6.5.3   Segregated land use

Segregated land use consists of large tracts of similar land use that requires use of the automobile for basic daily destinations (Figure 6.8c). Since mixed land use areas may look segregated at a micro-level, the definition of segregated land use employed here is building units that are located beyond reasonable walking distance to multiple other types of urban land uses. In order to accomplish this, the mix of land use is examined within a 1500 ft (457.2 m) pedestrian distance (the typical distance a pedestrian will walk in 10 minutes; Nelessen, 1995). Housing units within walking distance to multiple other types of urban land uses are considered *mixed*, while housing units with only other housing within the pedestrian distance are considered *segregated*.

The segregated land use metric was calculated by converting the vector-based 'urban' land use/land cover data layer to a grid. The dataset included 18 different classes of *urban* land use, some of which were recoded to better reflect the segregated land use analysis. A *neighbourhood variety* calculation was performed on the gridded urban land use, utilizing a radius of 1500 ft (457.2 m) to represent the pedestrian distance. This produced a grid surface where every cell was enumerated according to the variety or mixture of different urban land use categories within the search radius.

Since the other sprawl indicator measures produce output in which higher values indicate higher sprawl, the *mixed land use* surface grid was inverted to a *segregated land use* value, where higher numerical values represent a greater indication of the non-mixed (i.e. segregated) characteristic associated with sprawl. This was accomplished by subtracting the mixed-use grid from a constant grid with a value equal to 1 plus the most mixed grid cell occurrence (in the pilot study the maximum mixed land use occurrence was 7). The value of the segregated land use grid for a 1500 ft radius was then assigned to each housing unit point. The municipal-level segregated land use index (SL_{mun}) was calculated by averaging the segregated land use value of each new housing unit by municipality, as depicted in equation 6.3. New building units that have a higher segregated land use value are considered sprawling for this measure.

$$SL_{mun} = \frac{\sum Seg_{unit}}{\sum N_{unit}} \tag{6.3}$$

where SL_{mun} = segregated land use indicator by municipality, $Seg_{unit} = X$ – number of different developed land uses with 1500 feet (457.2 m), $X = 1$ plus the maximum land use mix in a given dataset (note: the baseline land use mix will vary by dataset), and N_{unit} = number of new residential units.

6.5.4 Highway strip

The highway strip development component of sprawl is usually typified by fast food restaurants and retail strip malls, but can also include single-family housing units lining rural highways (Figure 6.8d). However, this analysis focuses only on residential growth. As developed, the highway strip index is a binary measure. Residential units are designated highway strip if they occur along rural highways outside of town centres and the associated urban growth boundaries. New housing units within the delineated rural highway buffer are considered sprawling for this measure.

For this study, the highways were delineated from the dataset as all non-local roads (i.e. county-level highway or greater) outside of designated centres of the New Jersey State Plan. The buffer was set at 300 ft (100 m), a common depth for a 1 acre (0.405 ha) housing lot. Housing units that fell within the buffer were coded to 1 and units outside the buffer were coded to 0. The municipal level highway strip index (HS_{mun}) was calculated by summing the number of new residential units that occurred within the highway buffer and Normalizing by the total number of new units that were developed within the entire municipality, as depicted in equation 6.4. This provided, in essence, a probability measure of highway strip occurrence for each municipality. Municipalities that experienced a higher ratio of highway strip development were considered more sprawling for this measure than municipalities with lower ratios.

$$HS_{mun} = \frac{\sum HB_{unit}}{\sum N_{unit}} \tag{6.4}$$

where HS_{mun} = highway strip indicator by municipality, HB_{unit} = residential unit within the 300 ft highway buffer, and N_{unit} = number of new residential units.

6.5.5 Community node inaccessibility

The community node inaccessibility index measures the average distance of new housing units to a set of nearest community nodes (Figure 6.8e). The centres chosen in this analysis included schools, libraries, post offices, municipal halls, fire and ambulance buildings and grocery stores. The centres were chosen to reflect likely destinations for any residents within a community, as well as the availability of data for centre locations. The set of community nodes is intended to be an index, not an exhaustive set of destinations. It is argued that these selected destinations are reasonable proxy for destinations overall and thus provide valuable insight into the accessibility, as measured by road distance from each housing unit. Each selected community destination (i.e. node) was identified in the county-wide digital parcel map, utilizing the owner information as well as interpretation of digital orthophotos and hard-copy county maps.

New housing units were analysed for their road network distance to the community nodes, utilizing a cost/distance calculation over a gridded roads and urban mask. Road network distances were generated for each individual selected community node type to all housing units. The individual community node distance values were averaged into a single community node distance value. The municipal-level community node inaccessibility index (CNI_{mun}) was calculated by summarizing the new housing unit community node distance values by municipality as depicted in equation 6.5. Sprawling land use patterns have significantly higher average road distance between new units and the set of selected community nodes.

$$CNI_{mun} = \frac{\sum \overline{Dcn}_{unit}}{\sum N_{unit}} \qquad (6.5)$$

where CNI_{mun} = community node inaccessibility index by municipality, $\overline{Dcn}_{unit}$ = average distance of new residential unit to the set of community nodes, and N_{unit} = number of new residential units.

6.5.6 Normalizing municipal sprawl indicator measures

Each of the five individual sprawl metrics highlighted here reflects a particular geospatial characteristic of urban growth and provides useful analytical information. However, the measures are not standardized, but reflect an appropriate measurement unit for each particular trait. For example, some measurements such as *leapfrog* are linear distances, some such as *density* are areal measures and yet others such as *segregated land use* are in numbers of land uses. The diversity and range between these measurement units precludes direct comparison between metrics. Normalization of the measures through percentile rank, however, results in index values that can be cross-compared. Once the individual sprawl measures were normalized to percentage ranks, they were summed together to produce a single cumulative summary measure of sprawl, or what Hasse and Lathrop characterize as a *meta-sprawl indicator* for each municipality. Housing unit-level calculations facilitate a new approach for rescaling data. While the authors demonstrate rescaling to the municipal level (an appropriate scale due to local zoning control in New

Jersey), summary sprawl measures could be calculated for any geographical extent of interest by summarizing the individual housing units by any desired geographical unit, such as census tract, county or metropolitan area.

This case study demonstrates that the development of a housing unit-level urban database promises to provide a more robust means of analysing urban form for characteristics of sprawl and smart growth than previous urban data models. However, the development of such building unit-level databases for extensive spatial areas is challenging. Most of the socio-economic data that is available for analysis is aggregated to larger geographic areas, such as a census block, commuter zone or zip code. Digital parcel maps still do not exist for many areas. Furthermore, identifying the location of individual housing units on a metropolitan scale is a formidable task, resulting in large databases of potentially hundreds of thousands of records. Techniques of data compression, indexing and random sampling of housing-unit data may need to be developed in order to make the data more manageable for larger spatial scales.

Nonetheless, the potential advantages of analysing urban form at its atomic level warrant the effort of developing building-unit based urban geospatial databases. An urban atomic database model also has the potential for innovative integration of remote sensing. Integration can be potentially facilitated in data development, data enhancement and data updating. For example, in data development, building-unit point location may be accomplished through integrating remote sensing imagery with automated address matching of a regional telephone directory. Points could be generated by the GIS address-matching geo-location algorithm and then adjusted for increased spatial accuracy by an automated remote sensing image recognition system. Traditionally, GIS data have been utilized as ancillary data within a remote sensing environment, such as overlaying roads and census tracts to enhance classification accuracies. The urban atomization model turns this relationship around, where the point location is enhanced by remotely sensed data as ancillary information. The possibilities for integrating remote sensing with GIS through an urban atomization approach extend well beyond the analysis of sprawl. Nonetheless, urban atomization for sprawl analysis, in particular, holds significant potential for advancing the delineation, characterization and analysis of the phenomenon of sprawl at the elemental scale at which it occurs, one house at a time.

6.6 Future benefits of integrating remote sensing and GIS in sprawl research

The interest in sprawl from many stakeholders and agencies will continue to grow, due to the broad implications that continued patterns of sprawl will have for ecology, society, economics and politics. While there has been substantial advancement in the identification, characterization and analysis of sprawl over the past several decades, the research is still arguably in an early stage. This chapter has highlighted some of the ways in which the geospatial technologies of remote sensing and GIS are being utilized to study the phenomenon of sprawl on multiple levels, from the metropolitan level down to the building-unit level. The integration of remote sensing and GIS is both advancing and being advanced through this sprawl research.

The building unit-level analysis as highlighted in the second half of this chapter holds particular promise for benefiting from the joining of GIS and remote sensing, because it allows for new avenues of integration between the physical land cover information that remote sensing imagery can provide and the socio-economic information that is more readily available for GIS. A building unit-level integration of GIS and remote sensing is not only of interest from an academic perspective but also from a policy perspective, because it performs at a level that can provide meaningful information to the stakeholders of the urbanization process.

Ultimately, this is where geospatial research can make its greatest contribution to the understanding and management of sprawl. The integration of remote sensing and GIS can assist in developing sprawl analytical methods that are employable to academics, policy makers and multiple other stakeholders. By integrating the two platforms, the combined strengths of each can overcome a number of limitations of utilizing remote sensing or GIS separately. Integration will lead to progress in urban research in areas such as image recognition, object-orientated urban feature modelling and near-real-time land data updating. Furthermore, this research can lead to development of a better urban typological system that objectively and justifiably characterizes urbanization patterns into appropriate categories, based on specific goals of public interest, such as land use efficiency, transportation, water quality and environmental health.

Considering growing population pressures, the continuing pace of urbanization and the impacts associated with modern patterns of sprawl, the need to study sprawl will continue for the foreseeable future. The integration of remote sensing technologies and GIS will play a significant role in advancing the understanding of the phenomenon of sprawl, while hopefully providing the tools for steering urbanization towards less problematic forms.

References

Allen, E. (2001) INDEX: Software for community indicators. In *Planning Support Systems: Integrating Geographic Information Systems, Models, and Visualization Tools*. ESRI Press: Redlands, CA, USA.

Burchell, R. and Naveed, S. (1999) The evolution of the sprawl debate in the United States. *West. Northwest* **5**, 137–160.

Burchell R. W., Shad, N. A., Listokin, D., Phillips, H., Seskin, S., Davis, J. S., Moore, T., Helton, D. and Gall, M. (1998) *The Costs of Sprawl – Revisited*. Transportation Research Board Report No. 39. National Academy Press: Washington, DC, USA.

Civco, D. L., Hurd, J. D., Wilson, E. H., Song, M. and Zhang, Z. (2002) A Comparison of land use and land cover change detection methods. ASPRS–ACSM Annual Conference and FIG XX11 Congress, Washington, DC, USA; 22–26.

Cova, T. J., Sutton, P. and Theobald, D. M. (2004) Exurban change detection in fire-prone areas with nighttime satellite imagery. *Photogrammetric Engineering and Remote Sensing* **70**, 1249–1257.

El Nassar, H. and Overberg, P. (2001) What you don't know about sprawl. *USA Today* 22 February, 1A, 6A-9A.

Epstein, J., Payne, K. and Kramer, E. (2002) Techniques for mapping suburban sprawl. *Photogrammetric Engineering and Remote Sensing* **68**, 913–918.

Ewing, R. (1997) Is Los Angeles-style sprawl desirable? *Journal of the American Planning Association* **63**, 107–126.

Ewing, R., Pendall, R. and Chen, D. (2002) *Measuring Sprawl and Its Impact.* Smart Growth America: Washington, DC, USA.

Forys, E. and Allen, C. R. (2005) The impacts of sprawl on biodiversity: the ant fauna of the lower Florida Keys. *Ecology and Society* **10**(1): 25 [online]: http://www. ecologyand-society.org/vol10/iss1/art25/

Fulton, W., Nguyen, M. and Harrison, A. (2001) *Who Sprawls Most: How Growth Patterns Differ Across the United States.* Centre on Urban and Metropolitan Policy, The Brookings Institute: Washington, DC, USA.

Galster, G., Hanlon, R., Wolman, H., Colman, S. and Freihage, J. (2001) Wrestling sprawl to the ground: defining and measuring an elusive concept. *Housing Policy Debate* **12**, 681–717.

Gustafson, E. J. (1998) Quantifying landscape spatial patterns: what is the state of the art? *Ecosystems* **1**, 143–156.

Hasse, J. E. (2002) Geospatial Indices of Urban Sprawl in New Jersey. Doctoral Dissertation, Rutgers University, New Brunswick, NJ, USA; 224 pp.

Hasse, J. E. (2004) A geospatial approach to measuring new development tracts for characteristics of sprawl. *Landscape Journal: Design, Planning and Management of the Land* **23**, 1–4.

Hasse, J. E. and Lathrop, R. G. (2003a) A housing unit-level approach to characterizing residential sprawl. *Photogrammetric Engineering and Remote Sensing* **69**, 1021–1029.

Hasse, J. E. and Lathrop, R. G. (2003b) Land resource impact indicators of urban sprawl. *Applied Geography* **23**, 159–175.

Hasse, J. E. and Lathrop, R. G. (2001) *Measuring Urban Growth in New Jersey.* Centre for Remote Sensing and Spatial Analysis, Rutgers University, New Brunswick, NJ [online]: http://www.crssa.rutgers.edu/projects/lc/urbangrowth/nj_urban_growth.pdf

Herold, M., Couclelis, H. and Clarke, K. C. (2005) The role of spatial metrics in the analysis and modelling of urban land use change. *Computers, Environment and Urban Systems* **29**, 369–399.

Hess, George R., Daley, Salinda S., Dennison, Becky K., Lubkin, Sharon R., McGuinn, Robert P., Morin, V. Z., Potter, K. M., Savage, R. E., Shelton, W. G., Snow C. M. and Wrege, B. M. (2001) Just what is sprawl, anyway? *Carolina Planning:a Journal of the University of North Carolina Department of City and Regional Planning* **26**, 11–26.

Hurd, J. D., Wilson E. H., Lammery S. G., and Civco, D. L. (2001) Characterization of forest fragmentation and urban sprawl using time sequential Landsat imagery. Proceedings of the ASPRS Annual Convention, St. Louis, MO, USA.

Jensen, R., Gatrell, J., Boulton, J., Harper, B. (2004) Using remote sensing and geographic information systems to study urban quality of life and urban forest amenities. *Ecology and*

Society **9**(5): 5 [online]: http://www.ecologyandsociety.org/vol9/iss5/art5/

Johnson, M. P. (2001) Environmental impacts of urban sprawl: a survey of the literature and proposed research agenda. *Environment and Planning A* **33**, 717–735.

Kasanko, M., Barredo, J. I., Lavalle, C., McCormick, N., Demicheli, L., Sagris, V. and Brezger, A. (2006) Are European cities becoming dispersed? A comparative analysis of 15 European urban areas. *Landscape and Urban Planning* **77**(1–2), 111–130.

Lathrop, R. G. (2004) Measuring land use change in New Jersey: land use update to Year 2000. CRSSA Technical Report No. 17–2004–1, Rutgers University, New Brunswick, NJ, USA [online]: http://crssa.rutgers.edu/projects/lc/reports/landuse_upd.pdf

Lopez, R. and Hynes, H. P. (2003) Sprawl in the 1990s: measurement, distribution and trends. *Urban Affairs Review* **38**, 325–355.

MacDonald, K. and Rudel, T. K. (2005) Sprawl and forest cover: what is the relationship? *Applied Geography* **25**, 67–79.

McGarigal, K. and Marks, B. J. (1995) *FRAGSTATS: Spatial Pattern Analysis Program for Quantifying Landscape Structure.* USDA Forest Service General Technical Report No. PNW-GTR-351. US Department of Agriculture: Pacific Northwest Research Station, Portland, OR, USA; 122 pp.

Malpezzi, S. (1999) Estimates of the measurement and determinants of urban sprawl in US metropolitan areas. Unpublished paper, Centre for Urban Land Economics, University of Wisconsin, Madison, WI, USA.

Mesev, V. (2005) Identification and characterization of urban building patterns using IKONOS imagery and point-based postal data. *Computers, Environment and Urban Systems* **29**, 541–557.

Natural Resources Conservation Service (NRCS) (2004) *National Resources Inventory 2002 Annual NRI Land Use*: http://www.nrcs.usda.gov/technical/land/nri02/landuse.pdf

Nelessen, A. C. (1993) *Visions for a New American Dream: Process, Principles, and an Ordinance to Plan and Design Small Communities.* Edwards Brothers: Ann Arbor, MI, USA.

Openshaw, S. (1984a) The modifiable areal unit problem. In *Concepts and Techniques in Modern Geography* 38. GeoBooks: Norwich, UK.

Openshaw, S. (1984b) Ecological fallacies and the analysis of areal census data. *Environment and Planning A* **16**, 17–31.

Song, Y. and Knaap, G-J. (2004a) Measuring urban form: is Portland winning the battle against urban sprawl? *Journal of the American Planning Association* **70**, 210–225.

Sutton, P. C. (2003) A scale-adjusted measure of 'urban sprawl' using nighttime satellite imagery. *Remote Sensing of Environment* **86**, 353–369.

Sudhira, H. S., Ramachandra, T. V. and Jagadish, K. S. (2004) Urban sprawl: metrics, dynamics and modelling using GIS. *International Journal of Applied Earth Observation and Geoinformation* **5**, 29–39.

Theobald, D. (2001) Land use dynamics beyond the American urban fringe. *Geographical Review* **91**, 544–564.

Theobald, D. M. (2004) Placing exurban land use change in a human modification framework. *Frontiers in Ecology and Environment* **2**, 139–144.

Thornton, L., Tyrawski, J., Kaplan, M., Tash, J., Hahn, E. and Cotterman, L. (2001) NJDEP land use land cover update 1986 to 1995, patterns of change. Proceedings of the Twenty-First Annual ESRI International User Conference, San Diego, CA, USA.

Torrens, P. and Alberti, M. (2000) Measuring sprawl. Unpublished paper No 27, Centre for Advanced Spatial Analysis, University College London, London, UK.

US EPA (2005) *Smart Growth: Environmental Protection Agency*: http://www.epa.gov/smartgrowth/index.htm

Weng, Q. (2001) A remote sensing–GIS evaluation of urban expansion and its impact on surface temperature in the Zhujiang Delta, China. *International Journal of Remote Sensing* **22**, 1999–2014.

Wolman, H., Galster, G., Hanson, R., Ratcliffe, M., Furdell, K. and Sarzynski, A. (2005) The fundamental challenge in measuring sprawl: which land should be considered? *The Professional Geographer* **57**, 94–105.

Yeh, A. G. O. and Li, X. (1998) Sustainable land development model for growth areas using GIS. *International Journal of Geographical Information Science* **12**, 169–189.

Yeh, A. G. O. and Li, X. (2001) Measurement and monitoring of urban sprawling: a rapidly growing region using entropy. *Photogrammetric Engineering and Remote Sensing* **67**, 83–90.

7
Remote sensing applications in urban socio-economic analysis

Changshan Wu

Department of Geography, University of Wisconsin at Milwaukee, WI, USA

7.1   Introduction

Recently, urbanization has become a global phenomenon because of unprecedented world population growth and rural-to-urban migration. Currently, approximately 50% of the world's people reside in urban areas, and it is estimated that by 2030 about 60% of the total population will be urban dwellers (United Nations, 2002). Associated with this dramatic urban growth and urban sprawl, the geographical extents of built-up areas are expanding rapidly. In the USA, for example, approximately 10 million hectares of non-federal rural lands have been developed for urban land uses during 1982–1997 (US Department of Agriculture, 2000). Due to this rapid urbanization, scientists and urban planners are facing many challenges, including the loss of greenfield sites and increment of brownfield sites, shortage of utilities and resources, aggravated traffic congestion, and non-point environmental pollution. These problems associated with urbanization are severely impacting quality of life and economic development in urban areas (Black, 1996). In order to address these problems, it is imperative to understand and monitor urban systems, especially urban land use changes, socio-economic activity patterns and underlying processes, and the dynamic interactions between human activities and urban physical environments.

Understanding and monitoring urban systems requires both reliable data sources and robust analytical methods (Yang, 2003). Traditionally, surveying and mapping

Integration of GIS and Remote Sensing   Edited by Victor Mesev
© 2007 John Wiley & Sons, Ltd.

methods have been the major approaches for obtaining urban information. These methods, however, are labour-intensive and cannot provide timely information. Geographic information systems, together with aerial photographs, bring new data sources and techniques for urban information collection, storage, management and analysis. Census data, as an example, have been widely applied in neighbourhood analysis, transportation planning, urban environmental justice evaluation and population segregation studies (Wong, 1996; Peng *et al.*, 1997; Sui, 1999). Other GIS data, such as land parcel inventory, employment information, crime location data and public health information, are regularly collected and applied in various urban socio-economic analyses. In addition to GIS data, aerial photographs are frequently considered as references for urban information acquisition, including road networks, land parcels and building information. Other remote sensing data and associated image processing technologies, however, have been more or less ignored in urban analysis (Carlson, 2003). One major concern of urban planners is the coarse spatial resolution of remotely sensed data. As an example, the spatial resolutions of most popular remote sensing imagery, such as Landsat TM and ETM$^+$, SPOT, MODIS and AVHRR, vary from 20 m to 1 km. It is almost impossible to identify a house or a road segment from imagery with such coarse resolutions. The other concern is the applicability of remote sensing technologies in urban analysis. Urban planners may not recognize the potential contributions of remote sensing in urban analysis and are not familiar with the advancement of remote sensing technologies (Carlson, 2003). This chapter provides a comprehensive review of remote sensing applications in urban analysis, with special attention to urban socio-economic studies. The structure of this chapter is as follows. Section 7.2 discusses the principles of urban socio-economic studies using remote sensing technologies, and summarizes these studies into two groups: (a) socio-economic information estimation; and (b) socio-economic activity modelling. These two groups of studies are extensively reviewed in sections 7.3 and 7.4, respectively. In particular, section 7.3 reviews the applications of remote sensing technologies in socio-economic information estimation, and section 7.4 discusses the integration of remote sensing and GIS information for socio-economic activity modelling. Through reviews of current studies, the advantages and limitations of applying remote sensing technologies in urban socio-economic studies are given in section 7.5. The chapter is concluded in section 7.6.

7.2 Principles of urban socio-economic studies using remote sensing technologies

Remote sensing technologies have been successfully applied for estimating physical environments of urban areas. In particular, urban land use land cover information and their change patterns have been identified using remote sensing technologies with reasonable accuracy (Ward *et al.*, 2000; Civco *et al.*, 2002; Liu and

Lathrop, 2002; Guindon *et al.*, 2004). Moreover, the biophysical composition of urban environments, including vegetation, impervious surface and soil, has been successfully generated following the vegetation–impervious surface–soil (V–I–S) model proposed by Ridd (1995). In particular, urban vegetation distribution has been estimated by Small (2001, 2002), using spectral mixture analysis models. In addition, urban impervious surface fraction has been estimated using a number of models, including spectral mixture analysis (Phinn *et al.*, 2002; Wu and Murray, 2003; Wu, 2004), artificial neural network (Flanagan and Civco, 2001) and regression tree analysis (Yang *et al.*, 2003). Moreover, urban surface temperature and the associated urban heat island effects have been studied extensively (Lo *et al.*, 1997; Weng *et al.*, 2004).

While urban physical environment studies provide valuable information, urban socio-economic activities and their underlying forces are more interesting for urban planning and management. Population segregation, especially its patterns and underlying forces, as an example, has been an important topic in urban economic and social studies (Farley and Frey, 1994). Therefore, scientists have recently attempted to relate remote sensing information to socio-economic activities. However, remote sensing, as a means of acquiring information about the physical environment, cannot be directly applied for estimating or modelling socio-economic activities. The relationship between remote sensing information and socio-economic activities may be conceptualized in the theory of social space (Lo, 1997). In detail, socio-economic activities, conceptualized as socio-cultural environment, represent people's understanding of and reactions to physical environments. Therefore, socio-economic activities may be closely related to physical environments, which have been successfully estimated from remote sensing imagery. The recent developments of socio-economic research using remote sensing technologies can be subdivided into group 1, socio-economic information estimation, and group 2, socio-economic activity modelling. In group 1, the objective is to estimate socio-economic information of interest through regressing implicit or explicit urban morphological (physical) factors generated from remotely sensed data (see Figure 7.1). Figure 7.1a illustrates the process of socio-economic information estimation directly from remotely sensed data. In this process, urban physical environments are utilized implicitly. Figure 7.1b shows the steps of estimating urban socio-economic characteristics through regressing urban physical information obtained from remote sensing imagery. In group 2, the estimated urban morphological environments are utilized as environmental factors, together with other environmental or socio-economic information in GIS format, to model socio-economic activities (see Figure 7.2). The following two sections review these two groups of studies. Section 7.3 reviews socio-economic information estimation using remote sensing imagery, and section 7.4 reviews socio-economic activity modelling through the integration of remote sensing information and other environmental and socio-economic information in GIS format.

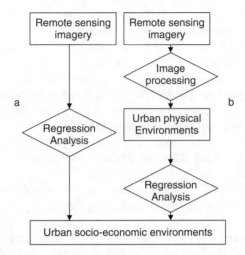

Figure 7.1 Remote sensing applications in urban socio-economic information estimation. (a) direct estimation from the radiance/reflectance of remote sensing imagery; (b) estimation with urban physical environments

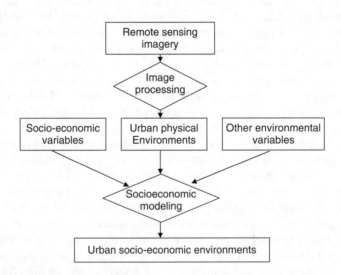

Figure 7.2 Remote sensing applications in urban socio-economic activity modelling

7.3   Socio-economic information estimation

There is a relatively long history of estimating selected socio-economic information from remote sensing imagery. Among these socio-economic characteristics, population is the most important socio-economic factor and has been extensively studied. In addition to population information, other socio-economic information, such as employment information, gross domestic product (GDP) and electrical power consumption, has been studied recently. In this section, the studies of socio-economic information estimations, including population, employment, GDP and electrical power consumption, are reviewed.

7.3.1   Population estimation

Population is an important socio-economic characteristic in urban studies because it is essential in supporting planning processes. In the design of public facilities, such as transportation infrastructure, libraries, public schools and hospitals, population distribution serves as a decisive factor. For example, a primary factor for transit route design is population density, because it is inadvisable to put a transit route in a region with low population density (Benn, 1995). In addition, population also plays an important role in private facility location analysis. In particular, population information is frequently utilized in retail store location site evaluation and insurance company customer analysis (Plane and Rogerson, 1994). Due to the importance of population information and the associated socio-economic attributes, censuses are taken regularly in most countries. Census population, however, is only available for every 5 or 10 years, although many planning activities require population information in non-census years. Moreover, for some developing countries, census data are not reliable or even available. Therefore, remote sensing provides an alternative means for estimating population information.

The earliest application of remote sensing in estimating population information utilizes the house-counting method applied to aerial photos (Lo, 1986a, 1986b). With this method, the number of houses is counted manually from the aerial photos. Then a survey is conducted to estimate the average number of persons per house. The total population for a study area can be calculated as a product of house numbers and average household size. Although this method proves to be relatively accurate, it involves manual interpretation of aerial photos. Time consuming and labour-intensive, this method is impractical for large urban areas. Moreover, accurate household size information, which is required for a variety of dwelling types, is difficult to obtain (Lo, 1989).

Automatic approaches with satellite remote sensing imagery, therefore, have been proposed for estimating population density (Lo, 1995). According to the information utilized for population estimation, these approaches can be classified into: (a) implicit estimation, in which information about spectral radiance/reflectance and their transformations is utilized for population estimation (see Figure 7.1a); and (b) explicit estimation, which utilizes urban physical parameters extracted from remote sensing imagery for population generation (see Figure 7.1b) (Cowen and Jensen, 1998; Jensen and Cowen, 1999). For implicit estimation, Lo (1995) utilized the radiances of bands 1, 2 and 3 of SPOT HRV imagery to obtain population density information in Hong Kong, China. In particular, he discovered that strong negative correlations existed between population density and the radiances of band 3 (0.79–0.89 μm) and band 1 (0.50–0.59 μm), and a positive correlation existed between population and the radiances of band 2 (0.61–0.68 μm). In his study, although urban biophysical parameters were not utilized, it was clear that the radiances of band 1 and 3 in SPOT HRV imagery were closely associated with vegetation concentration, which had a negative relationship with population density. Moreover, the radiance of band 2 was highly related to urban built-up areas; therefore, it had a positive correlation with population density. In addition to the radiances in individual bands, Harvey (2002a) utilized radiance transformations, including radiance squares, cross-product of radiances in different bands, radiance ratio from different bands, difference:sum ratio, etc. Moreover, Sutton et al. (1997) utilized light energy extracted from the Defense Meteorological Satellite Programme Operational Linescan System (DMSP-OLS) imagery for population estimation. The light energy represents the intensity of urban land uses, with higher energy existing in commercial and residential areas, and lower energy in agricultural areas. In summary, in these implicit estimations, although urban physical parameters are not directly utilized, specific urban physical environments are represented by radiances and their transformations. The other category of regression models utilizes urban biophysical and land use information extracted from remote sensing imagery for population estimation. As an example, urban land use types have been widely utilized. In particular, Lo (1995, 2003) utilized high and low land use areas to estimate zonal population counts. Chen (2002) applied three levels of residential density for projecting population density. Li and Weng (2005) also applied land use types as independent variables but developed separate regression models for different residential regions in Indianapolis, Indiana. In addition to land use information, Wu and Murray (2007) applied impervious surface fraction as an indicator of estimated population.

With information extracted from remote sensing imagery, it is necessary to construct a regression model to estimate population information. Reference population data, however, are only available at zonal levels (e.g. census tract), which are incompatible with the units of remote sensing information. Two types of modelling units, zone-based and pixel-based, have been utilized (Lo, 2003). The zone-based model involves aggregating pixel-based information to a zonal level, thereby performing regression analysis with the zonal data. This method has been popularly utilized for population estimation, and reasonable accuracies have been achieved (Sutton et al., 1997; Chen, 2002; Harvey, 2002a; Lo, 2003). This zone-based model also has two variations, one exploring the relationship between population counts and scale-dependent variables (e.g. area of a particular land use type), the other relating to the relationship between population density for a zone and a number of scale-independent variables (e.g. percentage of a particular land use type for a zone). In contrast to the zone-based models, the pixel-based method disaggregates zonal population to individual pixels (Harvey, 2002b). In particular, Harvey developed an expectation-maximization (EM) algorithm, and utilized iterative re-estimation approaches for population estimation. This model can be divided into an initial step and two iterative steps. In the initial step, the total population in a census zone is evenly redistributed to each residential pixel within that zone. Then two iterative steps are applied to refine the initial results and adjust the population count for each pixel until an acceptable result can be obtained. Wu and Murray (in press) compared these zone-based and pixel-based models for population estimation, and concluded that the pixel-based model provides slightly better results.

7.3.2 Employment estimation

Similar to population estimation, Lo (2004) explored the relationships between employment densities and apparent surface temperature extracted from Landsat 7 ETM$^+$ imagery in Atlanta, GA. In particular, Lo found that high surface temperature is highly correlated with high-density urban uses (commercial and industrial), which, in turn, are the geographic areas with high employment density. Based on this observation, Lo constructed a zone-based regression model to quantify the relationship between employments and surface temperature with census tract data (equation 7.1).

$$ln(\text{employment density}) = a_0 + a_1 \times ln(\text{surface temperature}) \qquad (7.1)$$

Results indicate that the correlation between employment density and surface temperature is statistically significant; Pearson's $R = 0.42$, $p < 0.01\%$.

7.3.3 GDP estimation

In addition to population and employment information, gross domestic product (GDP) has been estimated using remote sensing technologies (Elvidge *et al.*, 1997; Sutton and Costanza, 2002). In particular, the Defense Meteorological Satellite Programme Operational Linescan System (DMSP-OLS) was recently utilized for generating GDP estimates globally. The visible–near-infrared (VNIR) band (0.4–1.1 μm) of the DMSP-OLS has an ability to recognize low levels of VNIR radiance (e.g. commercial and residential light) at night. This light information in urban areas may provide some hints for urban economic activities, measured as GDP. One early application of GDP estimation was reported by Elvidge *et al.* (1997), who performed a log–log regression analysis for 21 American countries, including the USA, South American countries (e.g. Brazil, Colombia, etc.) and some island countries in America. The equation was reported as follows.

$$log(\text{GDP}) = -3.185 + 1.159 \times log \text{ (area lit)} \qquad (7.2)$$

The GDP is measured in billions of US dollars, and the area lit represents the light area identified from the DMSP-OLS imagery. Pearson's R^2 was reported as 0.97. With area lit as the independent variable, Doll *et al.* (2000) developed a similar regression model and applied it in 46 countries selected globally for GDP estimation. The regression model is:

$$log(\text{GDP}) = 0.9735 \times log \text{ (area lit)} \qquad (7.3)$$

The lit area for each country was obtained through analysing the DMSP imagery. The R^2 of the regression model is 0.85. In addition to the light area in the night, some other parameters have been utilized for GDP estimation. In particular, the total light energy measured from the DMSP-OLS imagery was utilized by Sutton and Costanza (2002) to estimate the GDP for global countries and Gross State Product (GSP) for every state in the conterminous USA. The formulation is as follows:

$$log \text{ (GDP)} = b_0 + b_1 \times log \text{ (light energy)} \qquad (7.4)$$

$$log \text{ (GSP)} = c_0 + c_1 \times log \text{ (light energy)} \qquad (7.5)$$

Moreover, Lo (2002) developed two new variables, light surface area and volume, for GDP estimation from the DMSP-OLS imagery. Specifically, a triangulated irregular network (TIN) was constructed based on the light radiance for every pixel. Then the light surface area and volume for each selected city were extracted from the TIN model. The results of this model are reported in equations 7.6 and 7.7:

$$log \text{ (GDP)} = d_0 + d_1 \times log \text{ (surface area)} \qquad (7.6)$$

$$log \text{ (GDP)} = e_0 + e_1 \times log \text{ (volume)} \qquad (7.7)$$

7.3.4   Electrical power consumption estimation

Together with GDP estimation, electric power consumption is another socio-economic characteristic that can be estimated from the DMSP-OLS imagery. With light area, light energy and light energy surface area and volume as independent variables extracted from the DMSP-OLS imagery, the electric power consumption has also been successfully estimated (Elvidge *et al.*, 1997; Lo 2002; Sutton and Costanza, 2002; Amaral *et al.*, 2005).

7.4 Socio-economic activity modelling

In addition to the direct estimation of socio-economic information, remote sensing technologies have been applied for socio-economic activity modelling. While the objective of socio-economic estimation is to generate specific information (e.g. population, employment) which is unavailable, socio-economic modelling assumes the availability of specific socio-economic information, and attempts to generate new information or discover the patterns and forces of socio-economic activities with the help of remote sensing imagery. In this chapter, several research areas involving the application of remote sensing information to socio-economic modelling are reviewed. In particular, section 7.4.1 explores the applications of remote sensing technologies in population interpolation; section 7.4.2 reviews the creation of socio-economic indices with the integration of remote sensing and other information; and finally, section 7.4.3 summarizes research related to understanding and modelling socio-economic phenomena, with a specific focus on housing price modelling.

7.4.1 Population interpolation

Population information is always collected and reported in enumeration zones, such as census blocks and tracts. This zone-based population data, however, creates problems in many geographic analyses (Martin, 1989, 1996; Fortheringham and Wong, 1991). One problem is related to data aggregation. Census data mask underlying individual population distribution because they are reported through aggregating population counts in pre-defined areal units. This aggregation requires an assumption of even population distribution within a zone for further geographical analysis (Moon and Farmer, 2001). Another problem is associated with data incompatibility (Bracken, 1993; Goodchild *et al.*, 1993). Varying zonal arrangements, such as school districts and transportation analysis zones, have been utilized by different departments and agencies for data collection and distribution. Therefore, a significant problem exists for data integration, which is needed for many geographical analyses and models. Further, the modifiable areal unit problem (MAUP) may exist when applying this zone-based population data in geographical analysis. In particular, the relationships between variables may only be a function of zonal arrangements, thus biased results may be obtained in statistical and spatial modelling (Openshaw, 1977; Martin, 1996).

One method of creating better population information is through the smart interpolation method, defined as transferring data with the help of additional information (Langford *et al.*, 1991; Harris and Longley, 2000). Remote sensing has served as important additional information for population interpolation. The studies by Langford *et al.* (1991) and Langford and Unwin (1994), in which land use and land cover data extracted from Landsat TM imagery is incorporated in generating better population information through regression analysis, are early applications of population interpolation. Similarly, Yuan *et al.* (1997) developed a generic linear model (GLM) to explore the relationship between population counts and land cover types generated from Landsat TM imagery. Further, they created a raster-based population surface (30 × 30 m) using the regression results and a scaling technique. In addition to regression technologies, Wu and Murray (2005) applied a co-kriging method to interpolate population density, by modelling the spatial correlation and cross-correlation of population counts and impervious surface fraction extracted from Landsat ETM$^+$ imagery. They proved that the co-kriging method is superior to regression techniques in exploring the relationship between population counts and remote sensing information. Moreover, they also performed a scaling method to remove other effects which cannot be modelled by remote sensing information. In addition to these population interpolation approaches, with which population is generated at fine resolutions and applied to small geographic areas, the Landscan Global Population Project (Dobson *et al.*, 2000) created a raster-based population surface at a resolution of 30 × 30 s and applied it globally. In this project, census counts are redistributed to raster cells based on additional information, including roads, slope, land cover, populated places, night-time lights, exclusive areas, urban densities and coastlines. Among these data, global land cover data were generated from the advanced very high resolution radiometry (AVHRR) imagery. The night light information was extracted from the DMSP-OLS imagery. With all of this information, the relative likelihood of population occurrence in each raster cell was calculated.

7.4.2 Socio-economic index generation

In addition to the applications in population interpolation, remote sensing information, together with other socio-economic information in GIS format, has been successfully applied in generating socio-economic indices (Lo, 1997). A residential quality index, for example, was developed with house size and vegetation percentage (including trees and grass) as positive indicators, and road and non-residential building densities as negative indicators (Forster, 1983). These indicators, such as house size and vegetation percentages, were extracted from Landsat TM imagery, aerial photographs, and housing sale records. As a further step, Weber and Hirsch (1992) developed three socio-economic indices, including a housing index, a quality index, and an attractivity index through integrating three data sources: SPOT XS imagery, census data and cartographic data. In particular, the housing index relates to housing size and the types of suburban houses. The housing index represents housing quality, with larger houses having higher values. Related to the housing index, the quality index was developed with housing quality and vegetation percentage as positive indicators, and high-density residential housing and commercial housing density as negative indicators. Finally, the attractivity index was formulated as positively related to the percentages of housing land cover and urban vegetation cover, and negatively related to industrial, commercial and parking land uses. Related to these studies, Lo (1997) integrated Landsat TM data with census socio-economic information to generate the quality-of-life index. In particular, environmental variables, including Normalized difference vegetation index (NDVI), surface temperature and percentage of urban land uses, were extracted from the Landsat TM imagery. Other socio-economic information, including population density, per capita income, median home value and percentage of college graduates, was obtained from census data. With all these data, Lo (1997) applied a principal component analysis to obtain the quality-of-life index. This index is positively correlated with per capita income, median home value, percentage of college graduates and NDVI, and negatively correlated with population density, percentage of urban land uses and surface temperature.

7.4.3 Understanding and modelling socio-economic phenomena

In addition to direct generation of socio-economic indices, remote sensing information has also been used to better understand and model socio-economic phenomena. One exciting application is the study of Weeks *et al.* (2004), in which remote sensing information was utilized to represent neighbourhood context for understanding fertility patterns and their changes during 1986–1996 in Cairo, Egypt. Specifically, the fractions of vegetation, impervious surface, soil and shade were generated from remote sensing imagery. Texture information was also extracted from the generated soil fraction image. These neighbourhood contexts, associated with social class/human capital variables and proximate determinants of fertility, were inputted to spatially filtered regression models. Results indicated that remote sensing information can significantly contribute to fertility pattern analyses in Cairo, Egypt. In addition to the fertility pattern analysis, Yu and Wu (2004) applied remote sensing information in understanding population segregation patterns and their environmental forces. In their studies, a population segregation index was calculated using census data. Remote sensing information, including fractions of vegetation, impervious surface, soil and land use types and texture information, was utilized to understand the influences of urban biophysical environments on urban population segregation phenomena. They concluded that remote sensing information makes significant contributions to understanding population segregation patterns. In addition to the above socio-economic modelling, remote sensing information also contributes to housing market analysis. Forster (1983) stated that remote sensing information can serve as environmental and locational parameters which may significantly affect housing prices. In particular, Forster (1983) utilized the housing density and road density extracted from Landsat data and aerial photographs, and the distance to the central business district (CBD), as independent variables to model house prices. In addition to Forster's (1983) work, Yu and Wu (2006) also incorporated remote sensing information to model house prices in Milwaukee, Wisconsin. In the study, they compared modelling results between a traditional hedonic model and the model with remote sensing information, and concluded that the addition of remote sensing information, in particular the product of soil and impervious surface fraction, can significantly improve housing model accuracy. In the following sections, studies integrating remote sensing and GIS information in population segregation analysis and housing value modelling are reported.

7.4.3.1 Population segregation analysis

Population segregation is considered to be an essential socio-cultural character-istic for analysing urban residential patterns. Traditionally, residential segregation patterns have only been analysed with census data, and environmental conditions have been ignored. In this study, urban environmental conditions were gener-ated from Landsat ETM$^+$ imagery applied to Milwaukee County, Wisconsin (see Figure 7.3), for analysing residential segregation patterns. In particular, biophys-ical parameters, including the fractions of vegetation, soil and impervious surfaces, were generated for each ETM$^+$ pixel using the normalized spectral mixture anal-ysis method proposed by Wu (2004). Fraction images (see Figure 7.4) show that high vegetation fraction is generally associated with suburb and rural areas, while

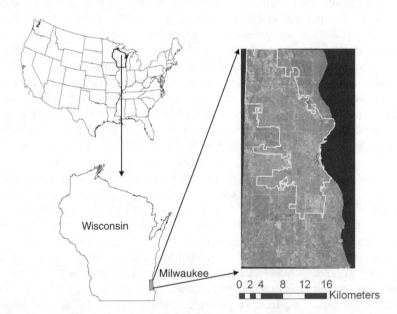

Figure 7.3 Location map of Milwaukee County, Wisconsin, USA. The right part shows an ETM$^+$ image acquired on 9 July 2001; the white line on the image illustrates the boundary of Milwaukee City

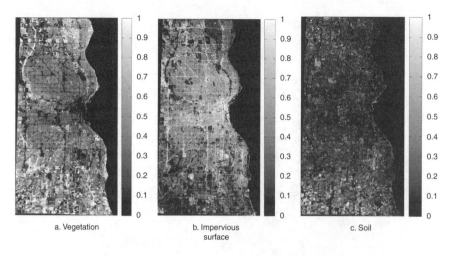

a. Vegetation   b. Impervious surface   c. Soil

Figure 7.4   Biophysical parameters generated from ETM$^+$ imagery

impervious surfaces are concentrated in commercial and high density residential areas. These biophysical parameters may represent environmental conditions, with highly vegetated and less-developed residential areas indicating desirable environmental quality. In addition to these three biophysical parameters, texture information was also extracted from the ETM$^+$ imagery. In particular, the *contrast* statistic was applied to represent urban environmental conditions. In this study, it is assumed that the areas with high contrast indicate low environmental quality. In addition to these urban environmental parameters, residential segregation information was quantified through local segregation indices. Although many segregation indices, such as the index of dissimilarity D (Duncan and Duncan, 1995), the spatial segregation index GD (Wong, 2005) and the entropy-based diversity index H (Plane and Rogerson, 1994), have been developed recently, they are essentially global measurements. Local segregation measures, indicating the relative degree of segregation for individual spatial units, are less discussed in the literature. The local segregation index (D_i) utilized in this study is similar to the local index developed by Wong (1996), with the addition of a scale adjustment factor. With 2000 census block group data in Milwaukee, the D_i can be calculated as follows:

$$D_i = \frac{m}{2} \left(\frac{b_i}{B} - \frac{w_i}{W} \right) \tag{7.8}$$

where D_i is the local segregation index for a particular block group i; b_i and w_i are African-American and White population counts in block group i; B and W are the total African-American and White population in the study area; and m is the total number of block groups (880 in the study area). The segregation index map

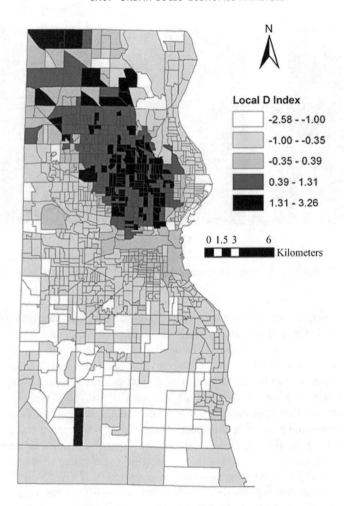

Figure 7.5 Local segregation index (D_i)

(see Figure 7.5) indicates a clear pattern of population segregation, with African-American population (higher D_i value) in the central to north-western part of the county, and White population (lower D_i value) in other areas.

With this segregation index (D_i) and the urban physical environmental parameters, global ordinary least squares (OLS) regression was applied to explore their relationships:

$$D_i = \alpha_0 + \alpha_1 \times High\_dens\_soil + \alpha_2 \times Low\_dens\_imp + \alpha_3 \times Contrast \quad (7.9)$$

where *High_dens_soil* indicates the fraction of soil in high-density residential areas; *Low_dens_imp* refers to the fraction of impervious surfaces in low-density

Table 7.1 Regression results between local segregation index and environmental parameters extracted from ETM$^+$ imagery

	Estimate	SE	Standardized coefficients	t Value	$p\ (>|t|)$
(Intercept)	−0.82	0.10		−8.00	4.34e − 15***
High_dens_soil	49.01	2.11	0.64	23.23	< 2e − 16***
Low_dens_imp	−3.13	0.35	−0.24	−9.00	< 2e − 16***
Contrast	0.02	0.005	0.10	3.56	0.0004***

Significance codes: *** = 0; ** = 0.01; * = 0.05.
Residual SE, 0.73 on 876 DF; F-statistic, 199.4 on 3 and 876 DF, $p < 2.2e − 16$; multiple R^2, 0.41; adjusted R^2, 0.40.

residential areas; *Contrast* is the contrast statistic extracted from ETM+ imagery, and α_0, α_1, α_2 and α_3 are regression coefficients. Regression results (see Table 7.1) suggest three findings. First, there is a significant relationship between residential segregation patterns and urban physical environmental information extracted from remotely sensed imagery. This indicates that remote sensing information is valuable in analysing segregation patterns. Secondly, the regression coefficients indicate that African-Americans are likely to live in high-density residential areas with high soil fractions, while White population concentration tends to occur in low-density residential areas with a high percentage of impervious surfaces. Moreover, African-Americans tend to live in areas with high spatial variations of land cover types. Third, the regression results also reveal that these environmental parameters can only account for approximately 40% of residential segregation pattern variations. This suggests the necessity of integrating other environmental or socio-cultural information for better explanations.

7.4.3.2 *Housing price modelling*

Besides its applications in socio-cultural studies (e.g. segregation), remote sensing information has also been applied in economic research. As an example, this section develops a specific application, in which remote sensing information is incorporated into a hedonic housing price model. The hedonic housing price model was initially developed by Rosen (1974), and has been widely applied as an econometric tool for exploring determinants of housing values. These determinants include housing structural attributes, location attributes and environmental conditions. Many hedonic housing models only include housing structural and/or location attributes, and the influences of environmental conditions on housing values are less studied. Therefore, one major objective of this study was to explore whether remote-sensing-generated environmental factors can contribute to housing price analyses.

The study area was the city of Milwaukee, Wisconsin, USA (see Figure 7.3). Three environmental parameters, including the fractions of vegetation, impervious

Table 7.2 Traditional hedonic housing price model with housing structural attributes

Coefficients	Estimate	SE	t Value	$p\ (>|t|)$
Intercept	8.345	1.184	7.048	0.000
Air conditioners	1.623	0.101	16.147	0.000
Floor size	0.056	0.181	0.310	0.756
Fireplaces	1.088	0.127	8.567	0.000
House age	0.368	0.073	5.014	0.000
Bathroom number	0.692	0.179	3.871	0.000
Story number	0.119	0.146	0.816	0.415

Dependent variable, house price, log-transformed.
Residual SE, 0.306 (training data), 0.274 (testing data).
Adjusted R^2, 0.689; F-statistic, 190.8 on 6 and 507 DF; $p < 2.2\mathrm{e} - 16$.

surface and soil, were derived from the ETM$^+$ image obtained on 9 July 2001. House price and structural attributes were obtained from the Master Property (MPROP) data file of the city of Milwaukee. First, a traditional hedonic model was developed with six selected housing structural variables, including floor size, number of bathrooms, number of stories, house age, and two dummy variables indicating whether or not air-conditioners and fireplaces were present. Specifically, a semi-log hedonic model was developed (equation 7.10):

$$\log P = \alpha_0 + \sum_{i=1}^{6} \alpha_i X_i \qquad (7.10)$$

where X_i indicates housing structural variables, and α_0 and α_i are regression coefficients. The results of this model are reported in Table 7.2.

In order to explore whether environmental attributes derived from remote sensing imagery can improve model results, a hedonic model with remote sensing information was constructed as follows:

$$\log P = \alpha_0 + \sum_{i=1}^{6} \alpha_i X_i + \sum_{j} \beta_j Y_j \qquad (7.11)$$

where Y_j represents environmental factors derived from remote sensing imagery, and β_j represents regression coefficients with these environmental factors. In this study, all three environmental factors (fractions of vegetation, impervious surface and soil) have significant contributions to the hedonic model. The product of impervious surface and soil fraction (SoilImp), however, is the most important factor, and thus was included in the model. The final result of this model is reported in Table 7.3.

Comparing Tables 7.2 and 7.3, three conclusions can be drawn. First, the relationship between housing values and the six housing structural attributes and remote sensing generated environmental factor (SoilImp) is significant. In both models, the

Table 7.3 Hedonic housing price model with housing structural attributes and environmental conditions generated from ETM$^+$ imagery

Coefficients	Estimate	SE	t Value	p $(>|t|)$
Intercept	10.476	1.131	9.262	0.000
Air conditioners	1.498	0.095	15.792	0.000
Floor size	−0.031	0.170	−0.184	0.854
Fireplaces	0.839	0.122	6.877	0.000
House age	0.090	0.076	1.189	0.235
Bathroom number	0.844	0.168	5.032	0.000
Story number	0.289	0.138	2.099	0.036
SoilImp	−5.626	0.643	−8.747	0.000

Dependent variable, house price, log-transformed.
Residual SE, 0.285 (training data), 0.263 (testing data).
Adjusted R^2, 0.730; F-statistic, 198.8 on 7 and 506 DF; $p < 2.2e − 16$.

adjusted R^2 indicate about 70% of variances of housing values can be explained by these hedonic models. Second, by comparing the adjusted R^2 of these two models, it can be found that the explanation power of the model improves about 4% through the addition of environmental attributes from remote sensing imagery. Moreover, a negative relationship between the housing value and the product of soil and impervious surface fraction has been obtained. This suggests that houses with lower values can be found in the regions where soil and impervious surfaces are abundant. Since the concentration of soil and impervious surface indicates deteriorated environmental conditions, the relationship is quite reasonable, and it proves that the environmental factors extracted from remotely sensed data should be incorporated in modelling housing prices.

Although in general the selected housing structural and environmental attributes can explain most of the variances in housing values, a close inspection of the coefficients and p values of several individual attributes reveals unexpected effects. In particular, the floor size, an important factor in housing value assessment, does not play a significant role in projecting housing values in both models. Moreover, in the model without remote sensing information, house age has a significant and positive influence on house values (see Table 7.2). However, this relation becomes insignificant when the environmental variable (SoilImp) is added (see Table 7.3). Further, in both models, the availabilities of air-conditioners and fireplaces are more important than other factors, such as number of bathrooms and bedrooms, and floor size. These results seem to be counter-intuitive. However, descriptive summary statistics (Table 7.4) indicate that the coefficients of variation (CV) of housing structural and environmental attributes vary significantly, with air conditioner availability (64.2%) and fireplace availability (126.7%) having the highest variations. This might partially explain why these two attributes are more important

Table 7.4 Descriptive summary statistics of housing structural attributes and environmental conditions (SoilImp)*

	Mean	SD	Coefficient of variation (%)	Min	Max
Air conditioners	0.372	0.239	64.2	0	0.879
Floor size (ft^2)	1358.9	417.3	30.7	877.8	5786.3
Fireplaces	0.15	0.19	126.7	0	0.96
House age (years)	74.3	23.6	31.8	13.33	122.50
Bathroom number	1.32	0.301	22.8	1.00	4.83
Story number	1.21	0.24	19.8	1	2.67
SoilImp	0.055	0.0239	45.5	0.004	0.128

*Average values at the census block group level are reported.

than others. Moreover, a correlation analysis (see Table 7.5) illustrates the existence of strong co-linearities among these housing structural and environmental attributes. In detail, the floor size is significantly and positively correlated with the number of stories ($R = 0.806$), fireplace availability ($R = 0.754$) and number of bathrooms ($R = 0.724$), while the house age is significantly and negatively correlated with the air conditioner availability ($R = -0.767$) and the product of the soil and impervious surface fractions ($R = -0.637$). The existence of such co-linearities among housing attributes might be the reason for the insignificance of floor size and the unexpected effects of housing age in the models. To address these problems, however, better modelling technologies must be developed.

Table 7.5 Correlation matrix among housing structural attributes and environmental conditions (SoilImp)

	Air conditioners	Floor size	Fireplaces	House age	Bathroom number	Story number	Soil-Imp
Air conditioners	1.000	−0.236	0.207	−0.767*	0.325	−0.182	0.418*
Floor size (ft^2)		1.000	0.754*	0.305	0.724*	0.806*	−0.182
Fireplaces			1.000	−0.180	0.825*	0.512*	0.120
House age (years)				1.000	−0.303	0.340*	−0.637*
Bathroom number					1.000	0.514*	0.180
Story number						1.000	−0.204
SoilImp							1.000

*Significant correlation at 0.01 level between the two crossed variables.

7.5   Advantages and limitations of remote sensing technologies in socio-economic applications

The recent developments of remote sensing technologies bring a potentially scientific basis for urban socio-economic applications. Remotely sensed technologies provide valuable information and innovative approaches for estimating the patterns of socio-economic activities and understanding their underlying forces. Unlike zone-based socio-economic information, such as census data, remotely sensed imagery provides self-consistent and objective measurements of urban physical environments (Miller and Small, 2003), which in turn reflect urban socio-economic activities. In the rest of this section, the advantages and limitations of remote sensing technologies in socio-economic information estimation and modelling are discussed.

7.5.1   Socio-economic information estimation

The advantages of applying remote sensing technologies to socio-economic information estimation can be summarized as follows. First, some socio-economic information, such as census data, cannot be obtained on a timely basis for the purposes of urban planning and management. Remote sensing imagery, however, can be obtained on a daily or monthly basis and thereby has the potential for providing updated socio-economic information. Second, for a few developing and less-developed countries, socio-economic information is unavailable or unreliable. Remote sensing imagery may be the only reliable resource for estimating socio-economic information, and may also be utilized for cross-validation. Finally, remote sensing technologies are very important for global estimations of socio-economic activities (e.g. population and GDP), because it is unlikely that such global socio-economic information can be obtained from other sources.

Although it is valuable to apply remote sensing technologies to socio-economic information estimation, there are still many limitations which prohibit further applications. One limitation is associated with the need for socio-economic information estimation. In developed countries, even many developing countries, detailed and updated socio-economic information is publicly available. Although some data are not collected in a timely manner, many methods have been applied to project the information for a particular period. Therefore, for developed countries, estimation of socio-economic information from remote sensing technologies may be unnecessary. The other limitation is associated with the estimation accuracy. Remote sensing is an important and precise tool for measuring urban biophysical and land use information, but its relation with socio-economic activities depends on many other socio-culture factors. Therefore, it is somewhat indirect and weak. Taking population estimation as an example, the reported population estimation accuracies for small areas were only approximate or less than 80%. Therefore, the applications of such estimated information may be problematic and create biased results in further urban planning and management.

7.5.2 Socio-economic information modelling

The integration of remote sensing and other sources of socio-economic and administrative data has great potential and has been successfully applied to urban land use and socio-economic activity modelling (Mesev *et al.*, 1995; Mesev, 1998a, 1998b; Miller and Small, 2003). Remotely sensed data can provide objective and timely information about urban physical environments, which cannot be easily obtained from other sources or *in situ* measurements. This information can be considered as an important input, which represents environmental or neighbourhood characteristics, to socio-economic models. Therefore, it is possible to explore the causal influences of urban physical environments on socio-economic activities. In addition, the detailed urban biophysical information extracted from remote sensing imagery is valuable in disaggregating zonal socio-economic data and generating better socio-economic indices. Overall, there is great potential for the integration of remote sensing and GIS technologies in socio-economic activity modelling.

However, there are still some challenges in the application of remote sensing technologies to socio-economic modelling. One challenge is the lack of communication between remote sensing researchers and social scientists. Currently, remote sensing researchers do not fully understand socio-economic processes and modelling methodologies, and social scientists do not acknowledge the need of remote sensing technologies and may not appreciate the benefits of these technologies. The other challenge is associated with the difficulties of incorporating remote sensing information into socio-economic modelling. For example, population counts for large areas (e.g. county) in a non-census year are typically estimated by demographic and economic approaches, in which population counts are calculated by taking the latest census enumeration and adding births and net migration and subtracting deaths (Bryan, 2000, 2004). For small areas (e.g. census tracts), population counts are estimated using the housing unit method (HU), in which electric bills, building permits and other administrative information are typically applied for estimation (Smith *et al.*, 2002). It may be possible to incorporate remote sensing information into these existing models, but further extensive investigations are needed.

7.6 Conclusions

This chapter discusses the principles of remote sensing applications in socio-economic studies, and summarizes these applications into two groups: (a) socio-economic information estimation; and (b) socio-economic activity modelling. In the first group, the estimates of several socio-economic characteristics, including population, employment, GDP and electric power consumption, have been reviewed.

For the second group, several applications, such as population interpolation, socio-economic index generation and socio-economic activity understanding and modelling, have been summarized. Finally, this chapter discusses the advantages and limitations of these applications, and argues for the great potential of integrating remote sensing and GIS for socio-economic modelling.

References

Amaral, S., Camara, G., Miguel, A., Monteiro, V., Quintanilha, J. A. and Elvidge, C. D. (2005) Estimating population and energy consumption in Brazilian Amazonia using DMSP night-time satellite data. *Computers, Environment and Urban Systems* **29**, 179–195.

Benn, H. P., (1995) Synthesis of transit practice 10: bus route evaluation standards. *Transit Cooperative Research Program, Transportation Research Board*. National Academy Press: Washington, DC, USA, 9–22.

Black, W. R. (1996) Sustainable transportation: a US perspective. *Journal of Transport Geography* **4**, 151–159.

Bracken, I. (1993) An extensive surface model database for population related information: concept and application. *Environment and Planning B* **20**, 13–27.

Bryan, T. (2000) US Census Bureau population estimates and evaluation with loss functions. *Statistics in Transition* **4**, 537–548.

Bryan, T. (2004) Population estimates. In Siegel, J. S. and Swanson, D. A. (eds), *Methods and Materials of Demography*. Elsevier/Academic Press: San, Diego, California, USA, 523–560.

Carlson, T. (2003) Application of remote sensing to urban problems. *Remote Sensing of Environment* **86**, 273–274.

Chen, K. (2002) An approach to linking remotely sensed data and areal census data. *International Journal of Remote Sensing* **23**: 37–48.

Civco, D. L., Hurd, J. D., Wilson, E. H., Arnold, C. L. and Prisloe, M. P. (2002) Quantifying and describing urbanizing landscapes in the Northeast United States. *Photogrammetric Engineering and Remote Sensing* **68**, 1083–1090.

Cowen, D. J. and Jensen, J. R. (1998) Extraction and modelling of urban attributes using remote sensing technology. In Liverman, D., Moran, E. F., Rindfuss, R. R. and Stern, P. C. (eds), *People and Pixels: Linking Remote Sensing and Social Science*. National Academy Press: Washington, DC, USA.

Dobson, J. E., Bright, E. A., Coleman, P. R., Durfee, R. C. and Worley, B. A. (2000) Landscan: a global population database for estimating populations at risk. *Photogrammetric Engineering and Remote Sensing* **66**, 849–857.

Doll, C. N. H., Muller, J., and Elvidge, C. D. (2000) Night-time imagery as a tool for global mapping of socioeconomic parameters and greenhouse gas emissions. *Ambio* **29**, 157–162.

Duncan, D. and Duncan, B. (1995) A methodological analysis of segregation indexes. *American Sociological Review* **20**, 210–217.

Elvidge, C. D., Baugh, K. E., Kihn, E. A., Kroehl, H. W., Davis, E. R. and Davis, C. W. (1997) Relation between satellite observed visible-near infrared emissions, population, economic activity and electric power consumption. *International Journal of Remote Sensing* **18**, 1373–1379.

Farley, R. and B. H. Frey (1994) Changes in the segregation of whites from blacks during the 1980s: small steps toward a more integrated society. *American Sociological Review* **59**, 23–45.

Flanagan, M. and Civco, D. L. (2001) Subpixel impervious surface mapping. Proceedings of the American Society for Photogrammetry and Remote Sensing Annual Convention, St. Louis, MO, USA, 23–27 April.

Forster, B. (1983) Some urban measurements from Landsat data. *Photogrammetric Engineering and Remote Sensing* **49**, 1693–1707.

Fotheringham, A. S. and Wong, D. W. S. (1991) The modifiable areal unit problem in multivariate statistical analysis. *Environmental and Planning A* **23**, 1025–1034.

Goodchild, M. F., Anselin, L. and Deichmann, U. (1993) A framework for the areal interpolation of socioeconomic data, *Environment and Planning A* **25**, 383–397.

Guindon, B., Zhang, Y. and Dillabaugh, C. (2004) Landsat urban mapping based on a combined spectral–spatial methodology. *Remote Sensing of Environment* **92**, 218–232.

Harris, R. J. and Longley, P. A. (2000) New data and approaches for urban analysis: modelling residential densities. *Transactions in GIS* **4**, 217–234.

Harvey, J. T. (2002a) Estimating census district populations from satellite imagery: some approaches and limitations. *International Journal of Remote Sensing* **23**, 2071–2095.

Harvey, J. T. (2002b) Population estimation models based on individual TM pixels. *Photogrammetric Engineering and Remote Sensing* **68**, 1181–1192.

Jensen, J. R. and Cowen, D. J. (1999) Remote sensing of urban/suburban infrastructure and socio-economic attributes. *Photogrammetric Engineering and Remote Sensing* **65**, 611–622.

Langford, M., Maguire, D. J., and Unwin, D. J. (1991) The areal interpolation problem: estimating population using remote sensing in a GIS framework. In Masser, I. and Blakemore, M. (eds). *Handling Geographical Information: Methodology and Potential Applications.* Longman: Harlow, UK, 55–77.

Langford, M. and Unwin, D. J. (1994) Generating and mapping population density surfaces within a geographical information system. *Cartographic Journal* **31**, 21–26.

Li, G. and Weng, Q. (2005) Using Landsat ETM$^+$ imagery to measure population density in Indianapolis, Indiana, USA. *Photogrammetric Engineering and Remote Sensing* **71**, 947–958.

Liu, X. and Lathrop, R. G. (2002) Urban change detection based on an artificial neural network. *International Journal of Remote Sensing* **23**, 2513–2518.

Lo, C. P. (1986a) Accuracy of population estimation from medium-scale aerial photography. *Photogrammetric Engineering and Remote Sensing* **52**, 1859–1869.

Lo, C. P. (1986b) *Applied Remote Sensing.* Longman: Harlow, UK.

Lo, C. P. (1989) A raster approach to population estimation using high-altitude aerial and space photographs. *Remote Sensing of Environment* **27**, 59–71.

Lo, C. P. (1995) Automated population and dwelling unit estimation from high-resolution satellite images: a GIS approach. *International Journal of Remote Sensing* **16**, 17–34.

Lo, C. P. (1997) Applications of Landsat TM data for quality of life assessment in an urban environment. *Computers, Environment, and Urban Systems* **21**, 259–276.

Lo, C. P., Quattrochi, D. A., and Luvall, J. C. (1997) Application of high-resolution thermal infrared remote sensing and GIS to assess the urban heat island effect. *International Journal of Remote Sensing* **18**, 287–304.

Lo, C. P. (2002) Urban indicators of China from radiance-calibrated digital DMSP-OLS nighttime images. *Annals of Association of American Geographers* **92**, 225–240.

Lo, C. P. (2003) Zone-based estimation of population and housing units from satellite-generated land use/land cover maps. In Mesev, V. (ed.). *Remotely Sensed Cities.* Taylor and Francis: London, UK.

Lo, C. P. (2004) Testing urban theories using remote sensing. *GIScience and Remote Sensing* **41**, 95–115.

Martin, D. (1989) Mapping population data from zone centroid locations. *Transactions – Institute of British Geographers* **14**, 90–97.

Martin, D. (1996) An assessment of surface and zonal models of population. *International Journal of Geographical Information Systems* **10**, 973–989.

Mesev, V. (1998a) Integration issues in GIS and remote sensing. *Computers, Environment, and Urban Systems* **23**, 1–3.

Mesev, V. (1998b) The use of census data in urban image classification. *Photogrammetric Engineering and Remote Sensing* **64**, 431–438.

Mesev, V., Longley, P. S., Batty, M. and Xie, Y. (1995) Morphology from imagery: detecting and measuring the density of urban land use. *Environmental Planning* **27**, 759–780.

Miller, R. B. and Small, C. (2003) Cities from space: potential applications of remote sensing in urban environmental research and policy. *Environmental Science and Policy* **6**, 129–137.

Moon, Z. K. and Farmer, F. L. (2001) Population density surface: a new approach to an old problem. *Society and Natural Resources* **14**, 39–49.

Openshaw, S. (1977) Optimal zoning systems for spatial interaction models. *Environment and Planning A* **9**, 169–184.

Peng, Z. R., Dueker, K. J., Strathman, J. and Hopper, J. (1997) A simultaneous route-level transit patronage model. *Transportation* **24**, 159–181.

Phinn, S., Stanford, M., Scarth, P., Murray, A. T. and Shyy, T. (2002) Monitoring the composition and form of urban environments based on the vegetation–impervious surface–soil (VIS) model by sub-pixel analysis techniques. *International Journal of Remote Sensing* **23**, 4131–4153.

Plane, D. A. and Rogerson, P. A. (1994) *The Geographical Analysis of Population with Applications to Business and Planning.* Wiley: New York, NY, USA.

Ridd, M. K. (1995) Exploring a V–I–S (vegetation–impervious surface–soil) model for urban ecosystem analysis through remote sensing: comparative anatomy of cities. *International Journal of Remote Sensing* **16**, 2165–2185.

Rosen S. (1974) Hedonic prices and implicit markets: product differentiation in pure competition. *Journal of Political Economy* **82**, 34–55.

Small, C. (2001) Estimation of urban vegetation abundance by spectral mixture analysis. *International Journal of Remote Sensing* **22**, 1305–1334.

Small, C. (2002) Multitemporal analysis of urban reflectance. *Remote Sensing of Environment* **81**, 427–442.

Smith, S. K., Nogle, J. and Cody, S. (2002) A regression approach to estimating the average number of persons per household. *Demography* **39**, 697–712.

Sui, D. Z. (1999) GIS, environmental equity, and the modifiable areal unit problem (MAUP). In M. Craglia and H. Onsrud (eds), *Geographic Information Research: Trans-Atlantic Perspectives.* Taylor and Francis: London, UK, 41–54.

Sutton, P., Roberts, D., Elvidge, C. and Meij, H. (1997) A comparison of nighttime satellite imagery and population density for the continental United States. *Photogrammetric Engineering and Remote Sensing* **63**, 1303–1313.

Sutton, P. C. and Costanza, R. (2002) Global estimates of market and non-market values derived from nighttime satellite imagery, land cover, and ecosystem service valuation. *Ecological Economics* **41**, 509–527.

United Nations (2002) *World Urbanization Prospects* (2002 revision). United Nations: New York, USA.

US Department of Agriculture (2000) *Summary Report: 1997 National Resources Inventory* (revised December 2000). Natural Resource Conservation Service, Washington, DC, and Statistical Laboratory, Iowa State University, Ames, IO, USA; 89 pages: http://www.nrcs.usda.gov/technical/NRI/1997/summary_report/report.pdf

Ward, D., Phinn, S. R. and Murray, A. T. (2000) Monitoring growth in rapidly urbanizing areas using remotely sensed data. *Professional Geographer* **52**, 371–386.

Weber, C. J. and Hirsch, J. (1992) Some urban measurements from SPOT data: urban life quality indices. *International Journal of Remote Sensing* **13**, 3251–3261.

Weeks, J. R., Getis, A., Hill, A. G., Gadalla, M. S. and Rashed, T. (2004) The fertility transition in Egypt: intraurban patterns in Cairo. *Annals of the Association of American Geographers* **94**, 74–93.

Weng, Q., Lu, D. and J. Schubring (2004) Estimation of land surface temperature-vegetation abundance relationship for urban heat island studies, *Remote Sensing of Environment* **89**, 467–483.

Wong, D. (1996) Enhancing segregation studies using GIS. *Computers, Environment and Urban Systems* **20**, 99–109.

Wong, D. (2005) Formulating a general spatial segregation measure. *Professional Geographer* **57**, 285–294.

Wu, C. (2004) Normalized spectral mixture analysis for monitoring urban composition using ETM+ imagery. *Remote Sensing of Environment* **93**, 480–492.

Wu, C. and Murray, A. T. (2003) Estimating impervious surface distribution by spectral mixture analysis. *Remote Sensing of Environment* **84**, 493–505.

Wu, C. and Murray, A. T. (2007) Population estimation using Landsat ETM+ imagery. *Geographical Analysis* **39**, 26–43.

Wu, C. and Murray, A. T. (2005) A cokriging method for estimating population density in urban areas. *Computers, Environment, and Urban Systems* **29**, 558–579.

Yang, L., Xian, G., Klaver, J. M. and Deal, B. (2003) Urban land-cover change detection through sub-pixel imperviousness mapping using remotely sensed data. *Photogrammetric Engineering and Remote Sensing* **69**, 1003–1010.

Yang, X. (2003) Remote sensing and GIS for urban analysis: an introduction. *Photogrammetric Engineering and Remote Sensing* **69**, 937–939.

Yu, D. and Wu, C. (2004) Understanding population segregation from Landsat ETM+ imagery: a geographically weighted regression approach. *GIScience and Remote Sensing* **41**, 145–164.

Yu, D. and Wu, C. (2006) Incorporating remote sensing information in modelling house values: a regression tree approach. *Photogrammetric Engineering and Remote Sensing* **72**, 129–138.

Yuan, Y., Smith, R. M. and Limp, W. F. (1997) Remodelling census population with spatial information from Landsat TM Imagery. *Computers, Environment, and Urban Systems* **21**, 245–258.

8

Integrating remote sensing, GIS and spatial modelling for sustainable urban growth management

Xiaojun Yang

Department of Geography, Florida State University, USA

8.1 Introduction

Urban areas are not only the engines of global economic growth but also magnets for new residents flooding in from rural areas (Knox and McCarthy, 2005). Over the past several decades, world-wide urban areas have experienced rapid growth in both human population and physical size. In 1950, only one-third of the world's 2.5 billion people were urban dwellers. In 2007, more than half of the 6.6 billion people of our planet live in cities. At the global scale, the growth of urban areas, or urbanization, shows no signs of slowing down and will likely continue unabated into the next two decades. While most non-American cities are more compact and clustered, cities in the USA have been experiencing a process of accelerated outward growth, with low-density suburbs spreading beyond the boundaries of central cities and over large areas of previously rural landscape (Kaplan *et al.* 2004). Such growth of suburbs, sparked largely by federal and state government policies, massive road building projects and automobile-dependent community planning, has considerably transformed the American landscape (Gillham, 2002). Urban growth has often been viewed as a sign of the vitality of a regional economy, but it has rarely been well planned, thus creating numerous environmental problems, such as accumulation of waste, air pollution, shortage of potable water, traffic congestion and loss of valuable farmland and green spaces among others (SDCN, 2001). These problems have placed an enormous burden on organizations responsible for the planning and management of urban areas.

For a long-term solution to the major environmental problems rising from urban development, synergies and strategic alliances between governments, local authorities, non-governmental organizations (NGOs), private enterprise and academia have been formed to promote urban sustainability, a movement that was adopted in the Agenda 21 document during the 1992 United Nations Rio Summit (UNCED, 1992). Sustainable urban development is widely defined as 'development that meets the needs of the present without compromising the ability of future generations to meet their own needs' (SDCN, 2001). Its ultimate goal is to optimize the economic and social conditions of citizens while respecting the need to preserve the world's environment and natural resources. It has become a vital imperative of international, national and local policies. The sustainable imperative has also made itself felt in planning and architecture. It has been fully incorporated into two interrelated perspectives on mitigating the urban sprawl that has become the predominant growth pattern in the USA: new urbanism and smart growth. The former was launched primarily by architects and physical planners, and focuses on physical form whose changes are considered as prerequisites for urban economic, social and ecological change (Knaap and Talen, 2005). Smart growth has been advocated predominately by environmentalists and policy planners, thus embracing more broad movements that satisfy growth while also minimizing the negative effects of urban sprawl (Gillham, 2002).

Sustainability goals involve the design and implementation of effective polices and plans to manage resources and provide services in the urban environment. This in turn requires accurate information bases and robust analytical technologies that can help evaluate urban growth dynamics and design environmentally sound development scenarios. Conventional survey and mapping methods cannot deliver the necessary information in a timely and cost-effective manner (Masser, 2001). Remote sensing, through sensors mounted on various air-borne or space-borne platforms, is providing the most important source of data for mapping the physical and cultural attributes of urban landscapes that can be used to monitor progressive urban development (Paulsson, 1992; Jensen and Cowen, 1999; Yang, 2002). GIS offers the power to integrate biophysical and socio-economic data, which can help us to understand the forces driving urban growth and development (LaGro and DeGloria, 1992; Foresman et al., 1997; Nedovié-Budié 2002; Burgi et al., 2004). Spatial modelling technologies depict complex structures of objects or events using mathematical equations that can be used to explore the inherent dynamics of complex urban systems and assess different scenarios for future urban growth (Alberti, 1999; Clarke et al., 2002; Cheng and Masser, 2003; Yang and Lo, 2003; Yeh and Li, 2003; Pijanowski et al., 2005). The integration of remote sensing, GIS and dynamic modelling technologies forms a platform for data acquisition, integration and modelling to support sustainable urban planning and management (Yang, 2000). Nevertheless, the urban environment, because of its complex and highly dynamic landscape, has been challenging the applicability and robustness of these methods and technologies (Longley, 2002). Further research efforts will certainly be maintained and will probably intensify in order to adapt these technologies to solve urban problems in a productive manner, thus reinforcing the absolute and comparative utility of current spatial information technologies.

The current research explores the applicability of integrating remote sensing, GIS and dynamic spatial modelling technologies to support sustainable urban growth management, using the city of Atlanta, Georgia, as a case. For the past three decades, Atlanta has been one of the fastest growing metropolises in the USA as it emerged to become the premier commercial, industrial and transportation urban centre of the south-east. The population increased 27% during 1970–1980, 33% during 1980–1990 and 40% during 1990–2000. The city has expanded greatly as suburbanization consumes large areas of agricultural and forest land adjacent to the city, pushing the peri-urban fringe farther and farther away from the original urban boundary. This rampant suburban sprawl has provoked concerns over losses of large areas of primary forests, inadvertent climate repercussions, and the degradation of the quality of life in this region (Bullard et al., 2000; Lo and Quattrochi, 2003). By using population trends, land use, traffic congestion and open space loss, Sierra Club's 1998 Annual Report ranked Atlanta as America's most sprawl-threatened large city (Sierra Club, 1998). Apparently, Atlanta has emerged as the sprawl capital of the nation and therefore is ideal for the study of urban sprawl. Since 1996, the author has been involved in various research projects aiming to understand the dynamics of change in Atlanta by using spatial information technologies. This article examines urban growth dynamics and future development scenario simulations, part of the above research effort. The primary objective of this work is to develop a comprehensive methodology for urban growth dynamics research that combines remote sensing, GIS and dynamic modelling for mutual reinforcement of the utility of these technologies. This should add useful insights to the emerging sustainable development management research, in which data integration and modelling of urban systems are central. Specifically, this project has three objectives: (a) to examine urban spatial growth and landscape change along the outskirts of the Atlanta region, as seen from a time series of satellite images; (b) to analyse the major forces driving the observed growth and change; and (c) to imagine, test and choose between two possible future urban growth scenarios under different environmental and development conditions.

8.2 Research methodology

A loose integration strategy was adopted in order to take full advantage of the sophisticated utilities available from various software packages. The research methodology

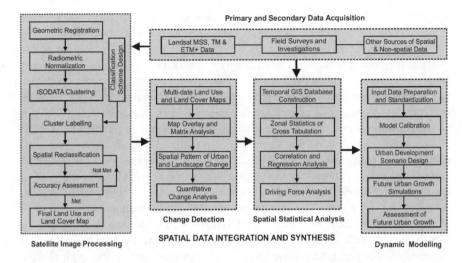

Figure 8.1 The flowchart of working procedural route that supported the monitoring of urban growth and landscape change, the analysis of forces driving the change and the prediction of future development scenarios. This technical framework comprised five major components: primary and secondary data acquisition; satellite image processing; change analysis; spatial statistical analysis; and dynamic modelling

can be divided into five phases: data acquisition and collection; satellite image processing; change detection; spatial statistical analysis; and dynamic modelling (Figure 8.1).

8.2.1 Study area

The geographic area of Atlanta specified here includes 10 counties under the Atlanta Regional Commission (ARC), as well as three additional counties, Coweta, Forsyth and Paulding, which have shown growth patterns similar to those of the ARC counties (Figure 8.2). The City of Atlanta is located in the centre. Physiographically, the Atlanta area is mainly in the foothills of the southern Appalachians in Northern Georgia, at an elevation of 300–350 m above sea level. The north-western part is in the Appalachian Mountains. It has an even terrain that slopes downward toward the east and south. The climate is generally characterized as mild. The Chattahoochee River traverses the study area.

8.2.2 Data acquisition and collection

This study used a time series of Landsat images as the primary data for measuring spatio-temporal urban growth at 6–8 year intervals, beginning in 1973. The Landsat

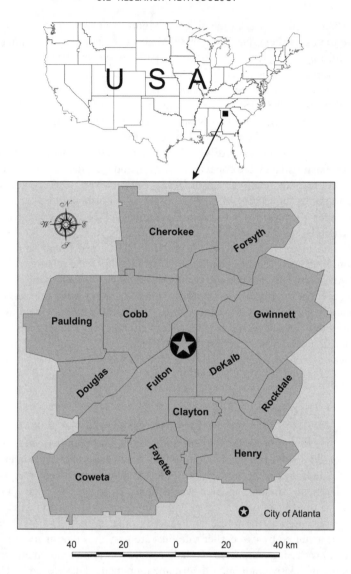

Figure 8.2 Location of the study area. It consists of the 10 counties under the Atlanta Regional Commission (ARC) and three additional counties, Forsyth, Paulding and Coweta, that show a similar growth pattern. The city of Atlanta is shown

MSS data were used for the period before 1982, when TM data are not available. After that period, TM and ETM$^+$ data were used. Eight predominantly cloud-free Landsat images were acquired by three sensors for Atlanta during 1973–1999 (Table 8.1). Most of the scenes were acquired during the spring or early summer seasons, when vegetation is in the stage of vigorous growth. The 1998 and 1999 scenes are the two exceptions. The 1998 TM scene, acquired in the winter season, was used for improving landscape mapping, particularly for identifying vegetation. The ETM$^+$ scenes were acquired in late summer because these were the only scenes free from clouds available between April and September 1999. Because of high image quality, the 1988 MSS scene was used as the reference image for relative radiometric normalization of other MSS data and no classification was attempted for this scene because a higher resolution scene for 1987 was available.

To facilitate change mapping, driving force analysis and urban growth simulation, a variety of existing data were collected, which include: digital images of Advanced Thermal and Land Applications Sensor (ATLAS) on 11 May 1997; contact prints of aerial photographs for 1986–1988; 1993 USGS digital orthophotos; the 1988–1990 land-cover classification from Georgia Department of Natural Resources; USGS 7.5 min Digital Elevation Model (DEM); US census survey data for 1980, 1990 and 2000; and major road networks from the AND Global Database (Table 8.1). In addition, field surveys were conducted to help establish the relationship between image signals and ground conditions. Representative spectral patterns for each land category on satellite image(s) were selected, along with the aerial photos corresponding geographically to these image spectral patterns. Field work also helped to obtain first-hand information about urban sprawl throughout the study area, which is useful for understanding the dynamics of change. A Trimble GPS receiver was used for better positioning during the field surveys.

8.2.3 Satellite image processing

The image processing procedures identified here include preprocessing, classification, GIS-based spatial reclassification and accuracy assessment. Geometric rectification and radiometric normalization were attempted for image preprocessing. The georeferencing strategy adopted was actually an image-to-image registration. The 1997 TM image had been accurately georeferenced by SpaceImaging EOSAT. This image was used as the reference to rectify other scenes. The 1973 and 1979 MSS images acquired by two earlier satellites are very different in contrast, despite the identical processing conducted. To help restore a common radiometric response among them, the relative radiometric normalization method developed by Hall *et al.* (1991) was applied to the two MSS images by using the 1988 MSS scene as the reference. Radiometric normalization was not attempted with the Landsat TM images because they have been processed to high radiometric quality.

The 1973, 1979, 1987, 1993 and 1999 images were classified using a two-step unsupervised method. First, the iterative self-organizing data analysis (ISODATA) algorithm was used to identify spectral clusters from image data, excluding the thermal band for the TM and ETM$^+$ scenes. Then, the resultant clusters were assigned to one of the six land use and land cover classes: high-density urban use

Table 8.1 List of the major themes in the spatial database

Group	No.	Theme	Date or year	Source(s)	Format	Purposes*
1	1	MSS images	4/13/1973, 6/11/1979, 5/14/1988	USGS EROS Data Centre	Rasters	A
	2	TM images	6/29/1987, 7/31/1993, 7/29/1997, 1/02/1998			
	3	ETM+ image	9/09/1999			
2	4	High-density urban	1973, 1979, 1987, 1993, 1999	Classified from Group 1		B, C, D
	5	Low-density urban				
	6	Cultivated/exposed land				B, D
	7	Cropland/grassland				B
	8	Forest				
	9	Water				
3	10	Elevation	1970s	USGS 7.5 min DEM		C
	11	Slope		Computed from slope		C, D
	12	Hill-shaded relief				D
	13	Excluded areas	1973 and 1999	Miscellaneous		D
4	14	Total population	1973, 1987, 1999	Interpolation from 1980 and 1990 census data		C
	15	Per capita income				
5	16	Cities	1990	US Bureau of Census	Points	C
	17	Roads	1973, 1987, 1999, 2025	AND global database updated with satellite imagery (Group 1); 2025 ARC Transportation Plan	Lines	C, D
	18	Major highway exits	1973, 1987, 1999		Points	
	19	Large shopping malls	1973, 1987, 1999	ESRI database updated with satellite imagery (Group 1)	Polygons	
6	20	Census-tract boundary	1980, 1990, 2000	US Bureau of Census		B, C, D
	21	County boundary				

*A, land use and land cover classification; B, urban and landscape spatial pattern and change analysis; C, urban growth and landscape change driving force analysis; and D, urban growth simulation and scenario development.

(commercial and industrial buildings and large open transportation facilities); low-density urban use (predominantly residential); cultivated/exposed land (areas with sparse vegetation cover); cropland/grassland (grasses, other herbaceous vegetation and crops); forest land (coniferous, deciduous, and mixed forests); and water.

Reclassification procedures were used to reduce the two types of misclassification errors on the initial maps produced through the unsupervised classification. The first type consisted of boundary errors due to the occurrence of spectral mixing within a pixel, which has been suppressed by using a modal filter. The second type consisted of spectral confusion errors due to the spectral similarity among several land classes, which is inevitable for an image acquired with a broad-band sensor and tends to be more perceptible in an urban scene than in a rural one. Defining the spectral confusions involves the use of image spatial and contextual properties. To this end, an image interpretation method was employed, because it allows the combined use of spectral and spatial contents as well as human wisdom and experience, thus providing the most powerful means for spectral confusion identification. Image interpretation can be incorporated effectively into a digital classification procedure with the use of on-screen digitizing, multiple zooming, area of interest (AOI) functionality and other relevant GIS tools, such as overlaying and recoding. In addition, several image-processing programmes permit advanced tools for geoprocessing, through which some 'manual' operations can be implemented automatically. Spectrally confused clusters were first identified, and AOI layers were created by on-screen digitizing. The AOI layers served as masks for splitting confused clusters. Finally, GIS reclassification functionality was employed to recode the split clusters into correct land classes. This was an interactive process until acceptable accuracy was obtained.

Due to the limited availability of reference data, it is impossible to perform accuracy assessment for each map exhaustively. The accuracy assessment strategy adopted here was to assess each type of imagery covering the study area through a standard method (Congalton, 1991). Results revealed that the land use and land cover maps classified from the TM or ETM$^+$ images yielded slightly better accuracy than those from the MSS images. Overall, all these maps met the minimum 85% accuracy stipulated by the Anderson classification scheme (Anderson *et al.*, 1976), indicating that the image processing approach adopted here has been effective in producing compatible land use and land cover data over time, despite the differences in spatial, spectral and radiometric resolution of the three generations of Landsat data used in this project.

8.2.4   Change analysis

Change analysis focuses on urban growth and the nature of change. To analyse the urban expansion, the spatial distribution of two major urban classes was extracted from each map in the time series. The change in urban land was summarized by using the GIS minimum dominate overlay method, which allows the smallest amount of high-density or low-density urban use in 1973 to show up fully, while only the net addition in the following years in the time sequence is shown. In this way, the urban extent of five dates was summarized on one map. By assigning a unique colour to the net addition of each year on the combined map, the progressive growth of high-density or low-density urban land can be perceived. The statistical summary of urban addition for each period was also generated.

The nature of change was analysed by characterizing land conversions through a two-way cross-tabulation or matrix analysis that assigns a unique class for each coincidence in two input layers, thus capturing different combinations of change. This method was used to characterize the land conversions for the periods 1973–1987 and 1987–1999. Given the total number of land classes, there are 16 possible combinations for each period. Because this study focused on the conversion of forest, cropland/grassland or cultivated/exposed land to urban uses, only nine were selected for further analysis, while the others were merged into a single unit. The combinations of conversion are: cultivated/exposed land to high-density urban use (C1); cultivated/exposed land to low-density urban use (C2); cropland/grassland to high-density urban use (C3); cropland/grassland to low-density urban use (C4); forest to high-density urban use (C5); forest to low-density urban use (C6); and cultivated/exposed land or cropland/grassland or forest to water (three conversions here combined into one) (C7). C0 is used for all other combinations which are not considered here.

8.2.5 Spatial statistical analysis

The purpose of spatial statistical analysis was to analyse the forces driving the observed changes. This necessitated the incorporation of additional data into the analysis. A total of 12 data layers were prepared, which were grouped into five major categories (Table 8.1):

1. *Statistical boundaries.* The 1980 and 1990 county and census-tract boundaries for the study area were extracted from the 1992 and 1995 TIGER street centreline files.

2. *Land use and land cover data.* Four classes, viz. high-density urban, low-density urban, cropland/grassland and forest, for 1973, 1987 and 1999, were used here.

3. *Topographic measures.* Terrain elevation and slope were derived from USGS 7.5 min DEM.

4. *Total population and per capita income.* These two measures were estimated for 1973, 1987 and 1999 by using data from several census years. This estimation was based on the assumption that the human population and per capita income grew exponentially and that their rates of increase were constant within

two immediate census years. At the county level, the total population and per capita income for the three specific years were computed through exponential interpolation with data from the four census years, viz. 1970, 1980, 1990 and 2000. At the census-tract level, a dasymetric mapping method described by Langford and Unwin (1994) was adopted to harmonize the original datasets before applying exponential interpolation due to the change of tract boundaries through time. The 1980 and 1990 census-tract boundaries are quite different, as the latter includes more than 100 new tracts. Through dasymetric mapping, the original 1980 dataset was remodelled by using the 1979 low-density urban extent. The 1980 remodelled data were summarized with the 1990 census-tract boundary. This compensated for the difference in tract boundaries between the 1980 and 1990 data. For exponential interpolation, the growth rate for 1980–1990 was used for the periods 1970–1979 and 1990–1999 but adjusted with the overall rates of increase for the entire study area.

5. *Location measures.* Urban centre proximity, highway proximity, node point proximity and shopping mall proximity were generated for the analysis at the tract level. In doing so, a weighting buffer grid was created from urban centres, major highways, nodes and large shopping malls respectively. Then, each grid was converted into a binary image with 1 as the area within the buffers and 0 as the background.

Statistical analysis was conducted at three different levels of aggregation, the entire study area, county and census tract. For the entire area, simple visual analysis is adequate to reveal trends with 3 years of data for one single observation. At the county level, there are 13 observations for 38 variables. At the tract level, there are 444 observations for 38 variables. These variables include: population densities, population density changes, per capita income, and per capita income changes; mean elevation, mean slope, percentages of county or tract in urban centre, road, node, and shopping mall buffers; proportions in high-density urban use and low-density urban use, as well as their changes over time, all for 1973, 1987 and 1999. Simple correlation analysis was used to determine whether an association exists as well as the magnitude and direction of the significant association. For data at the tract level, a multivariate regression was also conducted to examine the relationship between two urban class proportions (and their changes) as dependent variables with a group of independent variables. Stepwise variable selection was employed to determine which variables to include in the final model. A total of 12 models were computed, which relate urban uses, terrain conditions, demographic and economic variables and location measures to one another.

8.2.6 Dynamic spatial modelling

This part of the research was built upon the SLEUTH urban growth model (Clarke and Gaydos, 1998). It is a cellular automaton model whose behaviour is controlled by the coefficients of diffusion, breed, spread, slope resistance, and road gravity. This model was chosen because it is dynamic and future-orientated, conforming to the essential requirement of urban growth simulation in this project. The behaviour rules guiding urban growth in the model consider not only the spatial properties of neighbouring cells but also existing urban spatial extent, transportation and terrain slope. The driving force analysis (sections 8.2.5 and 8.3.2) found that transportation and terrain conditions were significant factors driving landscape changes in Atlanta. These behavioural rules therefore have realistically accounted for the driving forces in the formation of edge cities in Atlanta. In this research, the SLEUTH model was used as a tool to imagine, test and choose between two possible future urban growth scenarios under different environmental and development conditions for Atlanta.

To implement the model, a set of input data were prepared:

1. *Urban extent.* A time series of urban extent layers was derived by combining the two urban classes of the land classification maps for 1973, 1979, 1987, 1993 and 1999.

2. *Roads.* The 'road' layers contain not only major road networks but also node points and large shopping malls. The major highways and node points were extracted from the AND global highway database and then updated with the 1973, 1987 and 1999 images. Three layers of large mall polygons were extracted from the above images. A weighting system was established for highways, nodes and malls, respectively. The layers of highways, nodes and shopping malls in the same year were combined to form a single 'road' layer. In this way, a 'road' layer was produced for 1973, 1987 and 1999, respectively. The 'road' layer for the year of 2025 was produced by overlaying the 1999 roads with the improved roadways and new roadways proposed for 2025 by the Georgia Department of Transportation.

3. *Excluded area for development.* Two layers of excluded areas were assembled. The first layer is a binary image, consisting of water extracted from 1973 MSS image overlaid with various types of public land. These areas were not allowed for urban development. This layer was primarily used for the model calibration. For future growth prediction, another layer was built, with probabilities of exclusion included. All excluded areas in the first layer were preserved and assigned a value of 100. Additionally, this layer contains three levels of buffer zones around major streams in the study area.

4. *Slope and shaded relief.* These were derived from USGS 7.5 min DEM. Each layer was resampled into 240 m in grid size.

The next step was to calibrate the model for determining the best value of each control coefficient previously mentioned. This involved the use of statistical measures to quantify the historical fit between the modelled and historical data. Due to time and computational resources constraints, the calibration was broken down into three phases. The coarse calibration was to block out the widest range for each control coefficient. The fine calibration was to narrow down the ranges to approximately 10 or less. The final calibration was to determine the best combination. One more step was conducted to determine the starting values used for future growth simulation. The final values were: diffusion (71), breed (10), spread (32), slope resistance (73) and road gravity (100).

Two possible planning scenarios for future urban development in Atlanta under different policies and environmental conditions were designed and simulated. The first scenario assumes that the factors for growth remain unchanged; thus, it may be termed 'continuation', and provides a benchmark for comparison with the alternative growth strategy. To implement this scenario, the values of growth control coefficients obtained from the model calibration were used as the starting values. The 1999 urban extent data were actually used. The second scenario considers a hybrid growth strategy in which both conventional suburban development and alternative growth efforts, such as smart growth and new urbanism, are addressed. This scenario also considers environmental conservation by limiting development around several predefined stream buffer zones. To implement these ideas, the starting values for five growth control coefficients used in the first scenario were changed to slow down the growth rate and to alter the growth pattern. The actual control coefficients used for the second scenario were: diffusion (100), breed (100), spread (15), slope resistance (40) and road gravity (200). The proposed transportation improvements and new additions were considered here. The simulation of future growth was from 2000 to 2050 for both scenarios.

8.3 Results and discussion

8.3.1 Urban growth

Based on Figure 8.3 (left), the spatial expansion of high-density urban use is clearly visible. In 1973, the high-density urban use was small, occupying only 2.85% of the total land area for Atlanta (Table 8.2). The outward spread is quite clear in the 1979 and 1987 patterns, following the major transportation routes. Between 1973 and 1979, the net addition was 8292 hectares, or a 27.90% increment. The net addition was 16 265 hectares, or a 42.79% increment, between 1979 and 1987. These additions were primarily concentrated in several inner counties, such as Fulton, Cobb, DeKalb and Clayton. Significant growth took place by 1993 and 1999, with net additions of 19 844 and 33 197 hectares, respectively. These new development areas were highly concentrated in the northern and northeastern areas. The linearly concentrated pattern became more multinucleated. The 1999 distribution shows further enhancement of this transition as the spread took place largely along transportation routes and around urban centres, particularly in some peripheral counties, such as Coweta, Cherokee, Forsyth, Paulding and

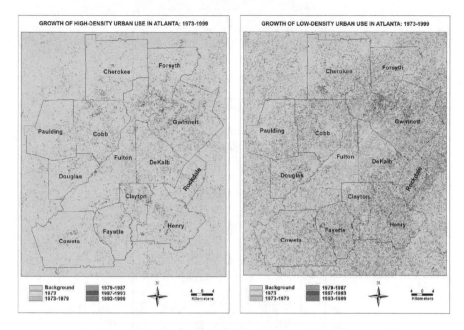

Figure 8.3 Spatial growth of the two major urban uses (Yang, 2002)

Fayette. In 1999, high-density urban use occupied 87 477 hectares, or 8.38% of the total land area, which is about a 194.31% increase in land compared with 1973. The daily increment was approximately 6 hectares or 15 acres between 1973 and 1999.

The evolution of spatial patterns of low-density urban use, mainly residential, is clearly perceived in Figure 8.3 (right). In 1973, low-density urban use occupied 76 910 hectares, or 7.36% of the total area. Although the low-density urban land shows signs of spreading outward, its large share was clearly concentrated in the inner city core and several inner counties. A somewhat linear pattern can also be seen along several major transportation highways. Thus, the spatial pattern of low-density urban use in 1973 was a form of concentration mixed with some degrees of dispersal. Significant growth occurred in 1979 and 1987, with net additions of 52 264 and 48 651 hectares, respectively (Table 8.2). Most of the new additions occurred outside the central city core, concentrating in four inner counties of DeKalb, Clayton, Fulton and Cobb, and in three exterior counties of Gwinnett, Rockdale and Fayette. Growth in the northern, north-western and north-eastern directions was quite clear. The spatial distribution pattern became more dispersed. Low-density urban use continued to grow after 1987. Most of the new development, however, took place in the exterior metropolis. This widely spread-out pattern is a major indicator of suburbanization. Quantitatively, low-density urban use occupied 282 959 hectares,

Table 8.2 Land use and land cover statistics for 13 metro counties in Atlanta

No.	Land use/cover	1973 Area (ha)	(%)	1979 Area (ha)	(%)	1987 Area (ha)	(%)	1993 Area (ha)	(%)	1999 Area (ha)	(%)
1	High-density urban	29722	2.85	38015	3.64	54280	5.2	67633	6.48	87477	8.3
2	Low-density urban	76910	7.36	129174	12.37	177825	17.03	214484	20.54	282959	27.1
3	Cultivated/exposed land	14534	1.39	20595	1.97	15511	1.49	21132	2.02	5358	0.51
4	Cropland/grassland	159345	15.26	117365	11.24	117686	11.27	96700	9.26	101122	9.68
5	Forest land	750366	71.85	724967	69.42	663673	63.55	625984	59.94	545148	52.2
6	Water	13404	1.28	14166	1.36	15306	1.47	18348	1.76	22217	2.13
	Total	1044281	100	1044281	100	1044281	100	1044281	100	1044281	100

or 27.10% of the total area in 1999, indicating a 267.91% increase between 1973 and 1999. The daily increment was about 22 hectares or 54 acres for the same period.

Table 8.2 also indicates the continuing decline in cropland/grassland and forest. The shrinking pattern of these two classes was proportional to the growth of the two urban classes. In general, the decline of cropland/grassland and forest land predominantly took place in the interior metropolis before 1987 but in the exterior after 1987. Quantitatively, cropland/grassland occupied 159 345 hectares, or 15.26% of the total study area, in 1973. It declined to 101 122 hectares (or 9.68%) by 1999. This represents a decrease of 36.54%, or a daily rate of 6 hectares (15 acres). Similarly, forest declined from 750 366 hectares (or 71.85%) in 1973 to 545 148 hectares (or 52.20%) in 1999, thus representing a decrease of 27.35%, or a daily rate of 22 hectares (53 acres) in land area.

The nature of the change is quite clear from Table 8.3. From 1973 to 1987, the loss of forest land contributed to the overwhelming share of the growth of the two major urban classes. High-density urban use had a net addition of 36 384 hectares, of which 62.17% came from the loss of forest land (C5) and 33.24% resulted from the loss of cropland/grassland (C3). The loss of cultivated/exposed land only contributed to 4.59% of the increase in high-density urban use (C1). For low-density urban use, 70.37% (C6) and 28.18% (C3) of the increase came from the loss of forest land and cropland/grassland, respectively. The loss of cultivated/exposed land only accounted for 1.45% (C1) of the net addition in low-density urban use. During 1987–1999, the magnitude of these conversions generally increased and forest and cropland/grassland conversions still overwhelmed the growth in two major urban uses.

8.3.2 Driving force

For the entire study area, the two urban classes have increased while the cropland/grassland and forest classes have declined since 1973. At the same time, population density and per capita income have rapidly increased, thus suggesting their impact on the urban growth and landscape changes (Table 8.4). The mean elevations for high-density urban use and forest have increased, while those for low-density urban use and cropland/grassland have decreased during the same period. This indicates that the land available for high-density urban development tends to be at higher elevations, while residential-dominated urban development favours relatively lower terrain. The mean slope gradient for each class tends to increase, suggesting that terrain slope gradient has been an influential resistance for urban development.

At finer spatial levels, the results of analysis are much more complicated as the landscape changes and demographic, economic and terrain characteristics have become more differentiated. The results of analysis for the two major urban classes are discussed below at the county and tract levels.

Table 8.3   Land use and land cover conversion statistics

Maps		Nature of change code*	Land conversion statistics			
Year A	Year B		1973–1987		1987–1999	
			Area (ha)	(%)	Area (ha)	(%)
3	1	C1	1669	0.16	3029	0.29
3	2	C2	1627	0.16	4709	0.45
4	1	C3	12094	1.16	11085	1.06
4	2	C4	31765	3.04	40127	3.84
5	1	C5	22621	2.17	28586	2.74
5	2	C6	79319	7.60	94559	9.05
3 or 4 or 5	6	C7	3653	0.35	6226	0.6
All other combinations		C0	891534	85.37	855960	81.86

*C1, converted from cultivated/exposed land (3) to high-density urban (1); C2, converted from culti-vated/exposed land (3) to low-density urban (1); C3, converted from cropland/grassland (4) to high-density urban (1); C4, converted from cropland/grassland (4) to low-density urban (2); C5, converted from forest (5) to high-density urban (1); C6, converted from forest (5) to low-density urban (2); C7, converted from cultivated/exposed land (3) or cropland/grassland (4) or forest (5) to water (6); and C0, all other combinations which are not considered here. Note that approximately 75% of total pixels remain unchanged.

8.3.2.1   High-density urban use

At the county level, the proportions of high-density urban use for 1973, 1987 and 1999 were positively correlated with population density, urban centre proximity, node proximity, mall proximity and highway proximity. The change in the high-density urban use proportion from 1973 and 1987 was positively correlated with population density change. Income and terrain factors did not exhibit a significant statistical relationship with either the high-density urban use proportion or its change through time.

At the census tract level, more delicate relationships were revealed. The proportions of high-density urban use for 1973, 1987 and 1999 correlated positively with road proximity, node proximity, urban centre proximity, population density and mean elevation, but negatively with mean slope gradient. The variables accounting for 45–49% of the variance in the 3 years' land class proportions often included road proximity, mean elevation, mean slope, per capita income and population density. The change in high-density urban use proportion during 1973–1999 correlated posi-tively with road proximity, node proximity, urban centre proximity, mean elevation and mall proximity, but negatively with mean slope. The independent variables that explained more than 53% of the variance in the class proportion change were

Table 8.4 Land use and land cover, demographic, economic, and topographic measures for the entire study area

Items		1973	1987	1999	1973–1987		1987–1999	
					Change	(%)	Change	(%)
Land use and land cover proportion (%)	High-density urban	2.846	5.198	8.377	2.352	82.626	3.179	61.159
	Low-density urban	7.365	17.028	27.096	9.664	131.212	10.068	59.122
	Cultivated/exposed land	1.392	1.485	0.513	0.094	6.722	−0.972	−65.457
	Cropland/grassland	15.259	11.270	9.683	−3.989	−26.144	−1.586	−14.075
	Forest	71.855	63.553	52.203	−8.302	−11.553	−11.350	−17.859
	Water	1.284	1.466	2.127	0.182	14.190	0.662	45.152
Per capita income (× 1000)		5.509	18.368	29.244	12.859	233.413	10.877	59.216
Population density (per hectare)		1.614	2.331	3.159	0.717	44.404	0.828	35.543
Mean elevation (100 ft)	High-density urban	9.578	9.626	9.663	0.048	0.504	0.037	0.381
	Low-density urban	9.516	9.484	9.432	−0.032	−0.340	−0.052	−0.549
	Cropland/grassland	9.653	9.563	9.463	−0.090	−0.931	−0.010	−1.041
	Forest	9.469	9.494	9.525	0.025	0.263	0.031	0.331
Mean slope (%)	High-density urban	4.862	4.986	5.557	0.124	2.550	0.571	11.452
	Low-density urban	5.509	5.812	6.220	0.303	5.500	0.408	7.020
	Cropland/grassland	5.081	5.327	5.277	0.246	4.842	−0.050	−0.939
	Forest	8.095	8.303	8.542	0.208	2.569	0.239	2.878

population density change, mean slope gradient, mean elevation, mall proximity, urban centre proximity and road proximity.

Statistical analysis indicates that population was more significant than location factors in explaining high-density urban use change at the county level. At the tract level, however, location and terrain conditions were clearly more significant than population. This is because commercial and industrial developments (the two major components of high-density urban use) were attracted to highways, node points and urban centres.

8.3.2.2   Low-density urban use

At the county level, the proportion of low-density urban use (mainly residential) for 1973, 1987 and 1999 correlated positively with population density. Other variables that correlated positively with low-density urban use proportion often included urban centre proximity, node proximity, mall proximity, highway proximity and per capita income. The mean slope gradient was found to correlate negatively with the proportion of low-density urban use in 1987 and 1999. The change in low-density urban use proportion during 1973–1999 correlated negatively with urban centre proximity, mall proximity, node proximity and highway proximity.

At the tract level, the proportion of low-density urban use for 1973, 1987 and 1999 correlated positively with population density, node proximity and highway proximity, but negatively with mean slope gradient. Other variables that correlated positively with the low-density urban use proportion for at least 1 year included per capita income and mall proximity. The variables accounting for 42–71% of the variance in the 3 years' class proportions included population density (3 years), mean slope gradient (3 years), road proximity (1973 and 1999), mall proximity (1987 and 1999), node proximity (1987) and mean elevation (1973). The change in the proportion of low-density urban use during 1973–1999 correlated positively with population density change but negatively with road proximity and node proximity. The independent variables that explained about 41% of the variability in the class proportion change during the same period were population density change, mean elevation, road proximity and mean slope gradient.

Population has been the most important factor for explaining changes in the proportion of low-density urban use at both the county and tract levels. At the tract level, it is very clear that highway proximity and node proximity were conducive to the rapid growth of low-density urban use, manifested by suburban housing. However, as more suburban housing has been built, its correlation with the location measures has become less important, thus explaining the decreasing value of the correlation coefficient with time, as well as the negative correlation between the proportion of 1973–1999 low-density urban use change and these location measures. It is also interesting to note that the proportion of low-density urban use correlated positively with per capita income for the years 1973 (0.62) and 1987 (0.16) only. The weakening correlation in more recent years indicates that at the beginning

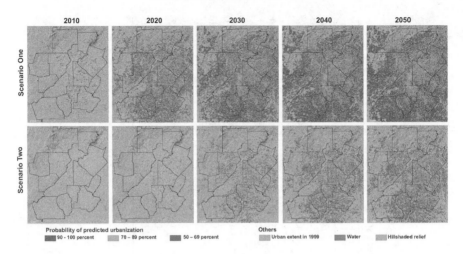

Figure 8.4 Simulation of future urban growth under two different scenarios. Note that the county boundary is overlaid

suburban housing was affordable only to people with higher income, but in recent years, as the general affluence of the population has increased, more people have been able to afford suburban housing.

8.3.3 Future growth scenario simulation

The progressive urban development as projected into the period 2000–2050 under two different scenarios can be well perceived from Figure 8.4. It is clear that a Los Angeles-like metropolis characterized by huge urban agglomerations would emerge by around 2030 if current development conditions were still valid (Scenario One). The vegetation area and open space in the 13 metro counties (excluding the north-western mountainous area) would be very limited. In contrast, the simulated urbanization under Scenario Two appears to be relatively restrictive, indicating that the effort of slowing down urbanization through model parameterization has been quite efficient.

Statistical measures reveal much more information (Table 8.5). Under Scenario One, the total urban area for 2050 would be 1 286 ,692 hectares. The total net increment in urban area with at least 50% probability would be 793 561 hectares, or 43.6 hectares/day on average, representing an increase of 160% between 1999 and 2050. The net urban increment as projected has been overwhelmingly concentrated within 1999–2030, further confirming the finding based on graphical outputs that a huge metropolis would take shape by 2030. As a result of such dramatic growth, urban land would occupy approximately 78.67% of the total modelled area by 2050. The average slope for urban land would increase from 4.87% in 1999 to 8.32%

Table 8.5 Statistics of simulation results for two different scenarios

		2010		2020		2030		2040		2050	
		Area (ha)	(%)	Area (ha)	(%)	Area (ha)	(%)	Area (ha)	(%)	Area (ha)	(%)
Scenario One	Probability										
	50–59**	41985	2.57	29889	1.83	21767	1.33	15978	0.98	12390	0.76
	60–69	51817	3.17	40602	2.48	29716	1.82	22067	1.35	17447	1.07
	70–79	60653	3.71	58343	3.57	43419	2.65	33615	2.06	26173	1.60
	80–89	61269	3.75	94654	5.79	76326	4.67	58291	3.56	46276	2.83
	90–100	27740	1.70	263820	16.13	469428	28.70	603821	36.92	691281	42.26
	Total urban area***	736595	45.03	980433	59.94	1133787	69.32	1226903	75.01	1286692	78.67
Scenario Two	Probability										
	50–59	29843	1.82	37123	2.27	31444	1.92	26346	1.61	22239	1.36
	60–69	22038	1.35	44444	2.72	41507	2.54	35983	2.20	31110	1.90
	70–79	7188	0.44	49144	3.00	54950	3.36	50970	3.12	45942	2.81
	80–89	559	0.03	43338	2.65	75635	4.62	80185	4.90	74920	4.58
	90–100	0	0.00	11048	0.68	75848	4.64	160865	9.83	238798	14.60
	Total urban area	552758	33.79	678223	41.46	772514	47.23	847480	51.81	906134	55.40

*It is computed by using urban area divided by the total modelled area (1 635 656 ha).

**This is the probability of predicted urbanization.

***It contains 1999 urban area (493 131 ha).

in 2050, implying that many steep, woody areas would be converted to urban use. Forest distribution in Atlanta was found to be positively correlated with terrain slope (section 4.2). Under Scenario Two, by 2050 the total urban area would be 906 134 hectares, or approximately 55.40% of the entire area. The total net urban increment would be 413 003 hectares, or 22.2 hectares/day, indicating an increase of 84% between 1999 and 2050. Apparently, the magnitude of urban growth as projected under this scenario has been substantially suppressed. The mean slope steepness for urban land would decrease from 4.87% in 1999 to 4.46% in 2050, implying that many low-lying, flat areas would be converted to urban uses.

The spatial distribution of simulated urbanization under the two scenarios can be discerned from Figure 8.4. For Scenario One, the projected urban additions for the period 1999–2010 are largely adjacent to the 1999 urban pixels, which can be viewed as a 'continuation' of urbanization. This corresponds to the fact that the overwhelming share of net urban additions under the first scenario was accounted for by organic growth. This type of urban growth actually represents the expansion of existing urban pixels into their surroundings. The projected urban additions during 2010–2030 are largely distributed over places far away from the 1999 urban land. Many projected additions are also found in western, north-western, and south-eastern parts. Some large urban clusters can be clearly recognized. The projected urban additions after 2030 are predominantly scattered over the western and south-eastern parts. Under Scenario Two, the projected urbanization for 1999–2010 is very limited. Most of the new additions are for the period 2010–2030, represented by blue and green pixels in Figure 8.4. Numerous large urban clusters can be clearly recognized, particularly in the southern and western parts.

Based on the above comparisons, it is found that the results of these two scenarios are quite different. Scenario One illustrates that unchecked urban sprawl would consume almost all of the vegetation and open space in the metropolitan area, with the exception of the north-western mountainous area. The dramatic growth in urban land, as projected under this scenario, would change the city's spatial form substantially, with numerous edge cities scattered over a huge area. This would greatly deteriorate the quality of life in Atlanta. Scenario Two would cut down the rate of urban growth by approximately 50%, compared to Scenario One. It would preserve much more greenness and open space, including predefined buffer zones along large rivers, streams and lakes. These preserved zones, although relatively small in area (about 27 358 hectares), contain the most important fresh water supplies, wetlands and floodplains for the metropolis. Therefore, Scenario Two should be most desirable for future growth planning in Atlanta.

8.4 Conclusions

Escalating urban growth throughout the world has provoked concerns over the degradation of our environment and the quality of life. How to optimize the development while minimizing its negative impacts becomes the ultimate goal of urban sustainability, a world-wide agenda that has been promoted as a vital imperative of international, national and local policies. Sustainable urban management involves the design and implementation of effective policies and plans to manage resources and provide services in cities. This in turn requires the introduction of innovative methodologies and technologies that can help understand urban growth dynamics and design environmentally sound development scenarios.

This study has demonstrated the utility of a loose-coupling approach to integrate remote sensing, GIS and dynamic spatial modelling for sustainable urban management. The entire research has gone through five major stages from the beginning of data acquisition and collection. The primary data were a time series of satellite images acquired by three Landsat sensors during 1973–1999. Other existing data, consisting mainly of census surveys, transportation and digital elevation models, were also collected, together with field surveys for ground truth acquisition. This was followed by image classification to produce a time series of land use and land cover maps from satellite data. Central to this work was the GIS-based image reclassification procedure, designated to resolve the spectral confusion. The land classification maps were then used to analyse urban growth and the nature of change. This was built upon the combined use of post-classification comparison and GIS techniques, which made possible the production of single-theme change maps, which emphasize spatial dynamics. The driving force analysis was conducted through a zonal-based approach to integrating biophysical data with socio-economic data from census surveys. The major challenge here has been to harmonize a variety of diverse data that differ in formats and projections, as well as parameter measuring and sampling methods. The adoption of a multilevel observation strategy was found to be useful for managing the MAUP. The last phase has been to incorporate several major driving forces into a dynamic urban model that was calibrated with historical urban extent data derived from the satellite series. The model was used as a tool to imagine, test and choose between two different scenarios under various environmental and development conditions. Based on this research, it is found that the integration of remote sensing, GIS and dynamic spatial modelling has mutually reinforced the utility of these techniques. The integration has provided insights that would not otherwise be available if spatial data were not organized in a GIS environment and GIS were not integrated with remote sensing and dynamic spatial modelling. Only through this integration can spatial information technology be effective for sustainable urban planning and management.

This study has established a well-documented regional case study focusing on Atlanta, well-known as America's leading sprawl city. This research reveals that every week, more than 100 acres of forest, green space and farmland were converted into urban uses in Atlanta. Between 1973 and 1999, the rate of urban expansion was 157% higher than that of the population growth, indicating the rapid and far-reaching urban sprawl in Atlanta. The driving force analysis reveals that counties with greater areas of urban growth tended to have larger demographic growth and higher socio-economic performance. Terrain and location conditions were more significant factors at the tract level than at the county level. Better location conditions have attracted more developments, and therefore urban growth and landscape changes have been intensified along major transportation corridors and around major urban centres. As more single-family housing units were built farther away from the large urban facilities, location conditions have become less correlated with housing development in recent years. The two scenarios designed with different environmental and development conditions have largely represented the major possible planning strategies for Atlanta. Scenario One simulated the continued growth trend if the urban sprawl is allowed to continue. Scenario Two simulated the development trend with a reduced rate and a different growth pattern. The modelling result suggests that Atlanta would be the next Los Angeles by approximately 2030 if the current growth rate and pattern do not alter. This will serve as a good warning to planners in Atlanta. In contrast, the result from the Scenario Two shows that much more greenness and open space, including buffer zones of large streams and lakes, could be preserved. Accordingly, Scenario Two should be the more desirable for the future urban growth of Atlanta. This suggests a smart growth strategy with emphasis on environmental protection, so that the 'livability' of the city of Atlanta will be maintained for future generations. These findings should be useful to those who need to manage resources and provide services to people living in this rapidly changing environment. Given that many major metropolises across the world face the growing problems caused by rapid urban development, the technical frameworks developed in the current study focusing on Atlanta should be applicable to other metropolises. This could improve our understanding of complex urban growth dynamics, thus facilitating a sophisticated approach to sustainable urban development.

References

Alberti, M. (1999) Modelling the urban ecosystem: a conceptual framework. *Environment and Planning B – Planning and Design* **26**, 605–630.

Anderson, J. R., Hardy, E. E., Roach, J. T. and Witmer, R. E. (1976) *A Land Use and Land Cover Classification System for Use with Remote Sensor Data.* USGS Professional Paper No. 964. USGS: Sioux Falls, SD, USA.

Bullard, R. D., Johnson, G. S. and Torres, A. O. (eds). (2000) *Sprawl City: Race, Politics, and Planning in Atlanta.* Island Press: Washington, DC, USA.

Burgi, M., Hersperger, A. M. and Schneeberger, N. (2004) Driving forces of landscape change – current and new directions. *Landscape Ecology* **19**, 857–868.

Cheng, J. Q. and Masser, I. (2003) Urban growth pattern modelling: a case study of Wuhan city, People's Republic of China. *Landscape and Urban Planning* **62**, 199–217.

Clarke, K. C. and Gaydos, J. (1998) Loose-coupling a cellular automaton model and GIS: long-term urban growth prediction for San Francisco and Washington/Baltimore. *International Journal of Geographic Information Science* **12**, 699–714.

Clarke, K. C., Parks, B. O. and Crane, M. P. (2002) *Geographic Information Systems and Environmental Modelling*. Prentice-Hall: Upper Saddle River, NJ, USA.

Congalton, R (1991) A review of assessing the accuracy of classification of remotely sensed data. *Remote Sensing of Environment* **37**, 35–46.

Foresman, T. W., Pickett, S. T. A. and Zippere, W. C. (1997) Methods for spatial and temporal land use and land cover assessment for urban ecosystems and application in the greater Baltimore–Chesapeake region. *Urban Ecosystems* **1**, 201–216.

Gillham, O. (2002) *The Limitless City: A Primer on the Urban Sprawl Debate*. Island Press: Washington, DC, USA.

Hall, F. G., Strebel, D. E., Nickeson, J. E., and Goetz, S. J. (1991) Radiometric rectification: toward a common radiometric response among multidate, multisensor images. *Remote Sensing of Environment* **35**, 11–27.

Jensen, J. R., and Cowen, D. C. (1999) Remote sensing of urban/suburban infrastructure and socio-economic attributes. *Photogrammetric Engineering and Remote Sensing* **65**, 611–622.

Kaplan, D. H., Wheeler, J. O. and Holloway, S. R. (2004) *Urban Geography*. Wiley: New York, NY, USA.

Knaap, G. and Talen, E. (2005) New urbanism and smart growth: a few words from the academy. *International Regional Science Review* **28**, 107–118.

Knox, P. L. and McCarthy, L. (2005)*Urbanization: An Introduction to Urban Geography*, 2nd edn. Prentice-Hall: Upper Saddle River, NJ, USA.

LaGro, J. A. and DeGloria, S. D. (1992) Land-use dynamics within an urbanizing nonmetropolitan county in New York State (USA). *Landscape Ecology* **7**, 275–289.

Langford, M. and Unwin, D. J. (1994) Generating and mapping population density surfaces within a geographical information system. *The Cartographic Journal* **31**: 21–26.

Lo, C. P. and Quattrochi, D. A. (2003) Land-use and land-cover change, urban heat island phenomenon, and health implications: a remote sensing approach. *Photogrammetric Engineering and Remote Sensing* **69**, 1053–1063.

Longley, P. A. (2002) Geographical information systems: will developments in urban remote sensing and GIS lead to 'better' urban geography? *Progress in Human Geography* **26**, 231–239.

Masser, I. (2001) Managing our urban future: the role of remote sensing and geographic information systems. *Habitat International* **25**, 503–512.

Nedovié-Budié, Z. (2002) Geographic information science implications for urban and regional planning. *URISA Journal* **12**, 82–93.

Paulsson, B. (1992) *Urban Applications of Satellite Sensing and GIS Analysis*. The World Bank: Washington, DC, USA.

Pijanowski, B. C., Pithadia, S., Shellito, B. A. and Alexandridis, K. (2005) Calibrating a neural network-based urban change model for two metropolitan areas of the Upper Midwest of the United States. *International Journal of Geographical Information Science* **19**, 197–215.

SDCN (Sustainable Development Communications Network) (2001) *Sustainable Cities: Environmentally Sustainable Urban Development*: http://www.rec.org/REC/Programs/SustainableCities/Introduction.html [accessed 19 June 2007].

Sierra Club (1998) *Sprawl: The Dark Side of the American Dream*: http://www.sierraclub.org/sprawl/report98/report.asp [accessed 19 June 2007].

UNCED (1992) Agenda 21: http://www.un.org/esa/sustdev/documents/agenda21/english/agenda21toc.htm [accessed 19 June 2007].

Yang, X. (2000) Integrating Image Analysis and Dynamic Spatial Modelling with GIS in A Rapidly Suburbanizing Environment. PhD Dissertation, University of Georgia, Athens, GA, USA.

Yang, X. (2002) Satellite monitoring of urban spatial growth in the Atlanta metropolitan region. *Photogrammetrical Engineering and Remote Sensing* **68**, 725–734.

Yang, X. and Lo, C. P. (2003) Modelling urban growth and landscape changes in the Atlanta metropolitan area. *International Journal of Geographical Information Science* **17**, 463–488.

Yeh, A. G. O. and Li, X. (2003) Simulation of development alternatives using neural networks, cellular automata, and GIS for urban planning. *Photogrammetric Engineering and Remote Sensing* **69**, 1043–1052.

9

An integrative GIS and remote sensing model for place-based urban vulnerability analysis

Tarek Rashed,* John Weeks,† Helen Couclelis‡ and Martin Herold§

*Department of Geography, University of Oklahoma, Norman, OK, USA

†Department of Geography, San Diego University, CA, USA

‡Department of Geography, University of California at Santa Barbara, CA, USA

§Institut für Geographie, Friedrich-Schiller-Universität Jena, Germany

9.1   Introduction

It is said that many centuries ago, an Indian princess asked the Buddha to summarize his philosophy for her. The wise man obliged, but when he brought his answer to the lady, she asked for a more concise summary. This exchange was repeated several times. Whenever the Buddha complied with her latest request, the princess kept on demanding an even shorter version. Eventually she asked: 'Can you express your philosophy in just *one* word?' Once more the Buddha obliged. The definition offered was 'Today' (Scheurer, 1994, p. 3).

At a glance, it appears impracticable in such a diverse and multidisciplinary area as urban vulnerability to environmental hazards to do what the Buddha did in philosophy – express the essence of the field in a single word. After all, six decades of considerable progress and outstanding achievements by hazards scholars have not succeeded in reconciling discrepancies surrounding fundamental concepts

Integration of GIS and Remote Sensing   Edited by Victor Mesev
© 2007 John Wiley & Sons, Ltd.

within the field (White and Haas, 1975; Mileti, 1999). The meaning of such basic terms as 'disaster', 'hazard', 'risk' and 'vulnerability' continues to be a matter of controversy (Dow, 1992; Cutter, 1996; Cardona, 2004). A review of the literature reveals considerable variation and fundamental conceptual differences among the numerous approaches and models developed to tackle vulnerability, risk and other hazard-related issues (Liverman, 1990; Dow, 1992; Cutter 1996; Rashed and Weeks, 2003; Cardona, 2004).

Despite all the controversies that exist in the field, we start this chapter with a proposition that urban vulnerability may indeed be summed up in one word – 'particularity'. As the literature suggests, the study of vulnerability is ecological in nature (Kates, 1971; Burton *et al.*, 1978; Andrews, 1985; Hewitt, 1997; Bolin and Stanford, 1999; Fitzpatrick and LaGory, 2000; Wisner *et al.*, 2004). As a result, an uneven and highly changeable complex web of dynamics and ecological factors, encompassing social, economic, cultural, political and physical variables, shape the patterns of urban vulnerability and determine the course in which these patterns evolve across space and through time. We refer to such context-dependent characteristics of vulnerability as 'particularity' to emphasize the notion that urban vulnerability can only be assessed in relation to a specific spatiotemporal context and its underlying dynamics, which interact together to produce particular forms of vulnerability.

We recognize that our attempt to describe the essence of vulnerability studies in one word is a bold step, especially when the reader is reminded that the word we use, 'particularity', has been central to philosophical tensions between various accounts of risks in hazards research (Mustafa, 2005). Accordingly, we do not expect the reader to accept our thesis as final. Rather, we invite the reader of this chapter to explore the plausibility of our thesis and its implications for the ongoing dialogue about the science of vulnerability (Cutter, 2001, 2003b) and the role of geographic information science and technology in risk and vulnerability analysis (Rejeski, 1993; Cova, 1999; Radke *et al.*, 2000; Cutter, 2003a).

The approach we pursue in our inquiry in this chapter is both theoretical and empirical. We first discuss epistemological positions on the particularity of urban vulnerability, drawn from contemporary work on hazards and disasters, to make the case for a place-based approach to vulnerability analysis. Next, we introduce the theoretical constructs of an integrative GIS and remote sensing model for place-based vulnerability analysis. We discuss how the proposed model could help resolve the dilemma of devising vulnerability assessments that recognize particularities in individual contexts, yet producing quantitative indicators to facilitate comparison of vulnerabilities across time and space. We then present a case study in which the model has been applied to assess the vulnerability of the metropolitan area of Los Angeles, California. We draw upon the results of this case study and conclude the chapter with a general discussion of integration issues in GIS and remote sensing technologies, and how such integration can provide a starting point for the science of vulnerability to evolve into a more robust field.

9.2 Analysis of urban vulnerability: what is it all about?

Vulnerability studies share in common the view that disasters are a product not only of hazardous events but also of social, economic and political environments. This is a crucial point indeed, as it puts vulnerability studies together under a unique theoretical paradigm that is quite distinct from other paradigms in disaster research, such as the technological-fix paradigm, which deems the geophysical processes that produce hazardous events to be more significant. The vulnerability approach to understanding urban disasters maintains the idea that calamities are poorly explained by the character of the events that may trigger them, be they natural (e.g. earthquake, flooding), technological (e.g. chemical release, dam failure), or caused by deliberate human action (e.g. terrorism act, war). Further, it asserts that the same damaging hazard could bring widely varying losses in societies, due to variations in social and physical vulnerabilities across urban places.

Despite the general conceptual ground they share, scholars of vulnerability are nonetheless divided amongst themselves on how to approach the question of vulnerability and the goals of its analysis. There have been several takes in the literature on the epistemological positions of vulnerability scholars (for recent reviews, see Wisner *et al.*, 2004, pp. 19–20; Mustafa, 2005, pp. 568–569). On the one hand, there is the realist view that emphasizes a set of common themes and elements to provide a better theoretical understanding of the 'real' root pressures in global, regional and national systems that shape the vulnerability profile of societies (Wisner *et al.*, 2004). Advocates of this view do not emphasize local particularities in their studies and consider doing so as a subtle form of environmental determinism. On the other hand, there are the pragmatist and constructivist views, which share a concern for the practicality of the context in which vulnerability is analysed, although they differ considerably in their methodological and philosophical foundations (Mustafa, 2005). For pragmatists, the emphasis on context particularities helps to introduce vulnerability analysis as a tool relevant to planners and decision makers. For constructivists, it provides a better means to comprehend the reality of disasters and to connect to local people.

Mustafa (2005) suggests that these above-mentioned epistemological differences regarding the understanding and analysis of vulnerability should not be seen as being in competition but rather as important complements. We concur with Mustafa's view and see it as a foundation upon which the recent idea that calls for a science of vulnerability (Cutter, 2003b) will need to rest. At one level, the concept of vulnerability in its broadest definition directs attention to the particular conditions that influence how well a society can cope with disasters and how rapid and complete its recovery can be. Findings of previous studies endorse the notion that these conditions do not come from 'outside' the urban place, neither do they erupt accidentally within it (Fitzpatrick and LaGory, 2000). Instead, they represent a product of everyday social life and ongoing urban dynamics that act upon the society and control its mutual relationship with the environment (Mitchell, 1989; Wisner, 1993; Cutter, 1996; Hewitt, 1997; Turner *et al.*, 2003; Tobin and Montz, 2004). At another level, there is a need to situate the finer detail brought about from examining local factors and particular patterns into a broader explanation of vulnerability, to gain deeper insights regarding the interdependence of vulnerability and differences between resources, societies and regions, and the interconnectedness among these groupings over space and time (Dow, 1992).

Reconciling the various epistemological positions on vulnerability into a more general analytical framework is therefore a central challenge to the emerging science of vulnerability and its role in 'help[ing] us understand those circumstances that put people and places at risk and those conditions that reduce the ability of people and places to respond to environmental threats' (Cutter, 2003b, p. 6). Our use of particularity as a keyword to summarize the essence of vulnerability analysis by no means negates the presence of a 'universal' knowledge of vulnerability, derived from important contributions by hazards scholars over the last two decades. The argument we make by using the 'particularity' keyword, however, is that for such knowledge to be effective in advancing risk-reduction goals, it is not enough to be credible (i.e. reasonably true and generally applicable). It also has to be salient (i.e. relevant to the needs of decision makers in a given context) and legitimate (i.e. not biased to a certain research culture) (ICSU, 2002). We argue that one path to create reliable, salient and legitimate knowledge of urban vulnerability lies in devising analytical approaches capable of acknowledging the contextual particularities of vulnerability while still allowing that knowledge to be transferred from one setting to another. In this chapter, we introduce one such approach and show the role that GIS and remote sensing can play in translating this place-based approach into a replicable methodology.

9.3   A conceptual framework for place-based analysis of urban vulnerability

As we have argued above, urban vulnerability is a place-dependent process residing in the 'socio-ecological' urban context; where 'social ecology' is a term used to emphasize the people–nature relationship (Andrews, 1985; ICSU, 2002). In order for such 'place-based' knowledge of vulnerability to be salient, it cannot be simply imported from the stock of universal knowledge (ICSU, 2002). It needs to be endogenously generated. Likewise, the socio-ecological contexts vary greatly between cities and even between neighborhoods within a given city. Consequently, the goals of urban vulnerability analysis (i.e. knowledge needs) are expected to vary too, to ensure legitimacy of the final product.

To illustrate the interrelationships between the place-based and universal levels of knowledge of vulnerability, and the way in which insights gained at local levels can

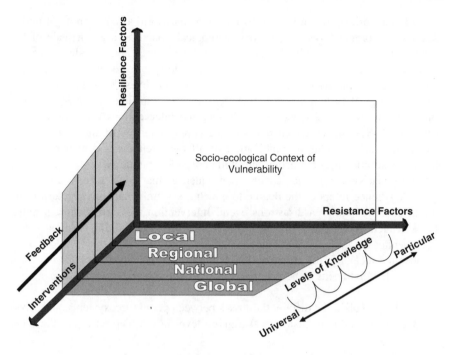

Figure 9.1 Simplified conceptual framework illustrating the interrelationships between the place-based and universal levels of knowledge of vulnerability

contribute to fundamental knowledge accumulated at the global level and vice versa, we present a simplified, general conceptual framework for vulnerability analysis in Figure 9.1. We have drawn on the insights of the vulnerability literature to establish the theoretical constructs of the proposed framework. We borrowed from Hewitt's ecological analysis of risk (Hewitt and Burton, 1971; Hewitt, 1997), Mitchell's contextual framework of hazards (Mitchell *et al.*, 1989), Cutter's hazards-of-place model (Cutter, 1996; Cutter *et al.*, 2000), and Mileti's systems approach to disasters (Mileti, 1999), the idea that patterns of vulnerability to hazards are contingent upon the physical, technological, social, economic and political realities of the system under consideration. We also have incorporated into the proposed framework some elements of Andrews' model of ecological risk intervention (Andrews, 1985) and Turner II *et al.*'s framework for vulnerability to climate change (Turner *et al.*, 2003), specifically the conception of urban areas as socio-ecological systems and the need to illuminate the nested scales of the vulnerability problem. Finally, we have used some elements of the 'pressure and release' model of vulnerability (Blaikie *et al.*, 1994; Wisner *et al.*, 2004) to convey the idea that locally focused studies and actions are of limited value if they do not account for the broader forces that affect the regional and local dynamics of vulnerability.

The framework shown in Figure 9.1 envisions the world as a hierarchy of multi-scale socio-ecological systems. A socio-ecological system at a given scale of the hierarchy encompasses the landscape of place(s) considered at this scale, i.e. neighborhood, city, region, country, as well as the people who reside in this landscape, their culture and the way in which they organize their lives. The vulnerability of a socio-ecological system at any hierarchical level is considered a collective function of the system's resistance, its resilience, and interventions measures applied at that level. System resistance refers to the coping capacity of the system *prior* to a disaster.It represents a combination of all the strengths and resources (e.g. physical, institutional, socio-economic, skilled personal, public awareness) available within a given system to face adverse consequences that could lead to a disaster. System resilience refers to the degree to which a system is capable to return to its normal conditions *after* a disastrous event. Intervention measures denote a range of risk reduction and mitigation measures applied to both building resilience and strengthening the system's resistance.

Generally speaking, the framework sets three main characteristics for the form of knowledge that needs to be generated from urban vulnerability analysis:

1. To help explain the differential losses between people, ecosystems, and physical features due to disasters at a given level in the hierarchy (i.e. the focal system).

2. To evaluate the ability of the focal system to absorb the impact of disasters (i.e. system resistance) while continuing to function and recover from losses (i.e. system resilience).

3. Ultimately, to determine the best options available to devise risk reduction measures.

The hierarchy in the framework has important implications on the forms of knowledge that could be generated, and consequently on the above-mentioned goals of vulnerability analysis.

First, the goals of vulnerability analysis, the problems it addresses and the factors and issues considered will vary by scale. What this means is that we cannot compare two systems, A and B, if they belong to different levels in urban hierarchy (i.e. if A represents a city and B represents a county). Second, the notion of hierarchy draws attention to the fact that any system in the hierarchy, whether large or small, is made up of smaller parts (a *suprasystem*) and at the same time is part of some larger whole of which it is a component (a *subsystem*). Consequently, understanding the vulnerability of a focal system (i.e. the level chosen to receive primary attention) requires the observer to attend both to the knowledge of vulnerabilities generated at the subsystems of that focal system and to the larger processes and dynamics operating at the suprasystem to which that focal system is related (Andrews 1985; Anderson *et al.*, 1999; Turner *et al.*, 2003). This means that one cannot compare the vulnerability of two focal systems, A and B, even though both are at the same level of hierarchy, unless they are part of the same suprasystem. For example, one may be able to compare the vulnerability of two cities belonging to Los Angeles County, California, but this comparison would be difficult if the cities belonged to different counties and if the processes found to be operating in these counties were different. It also means that the city A might be relatively more vulnerable than city B at one point of time and less vulnerable at another point of time, due to changes in the processes operating at the suprasystem to which they both belong.

Third, the hierarchy in the proposed framework views knowledge of vulnerability as a continuum from the particular to the universal and vice versa, as Mustafa (2005) has suggested regarding the complementary relationship among the epistemological positions in the field. As represented in Figure 9.1, the production of universal knowledge about vulnerability is accumulated and regularly updated through knowledge of vulnerability particularities generated at the lower levels of the hierarchy. These particular forms of knowledge at the lower levels are gradually generalized as we move to the upper levels in the hierarchy. In turn, the universal knowledge of vulnerability formulated at the upper levels is used to direct investigations into vulnerability conducted at lower levels. Finally, the proposed framework includes an axis for intervention measures that spans the hierarchy of socio-ecological systems. This axis emphasizes the idea that the goals of vulnerability analysis and decisions aiming at reducing risks are not quite the same across different scales in the hierarchy. At a regional scale, for example, decision makers may be concerned with the development of logistical and strategic plans to allocate resources. Therefore, it may be sufficient to crudely identify those areas that may experience higher degrees of damage in case of disasters. At the community level, on the other hand, it is necessary to have a thorough analysis of how the urban place will cope with a disaster to provide more specific intervention measures. Hence, the analysis would need to detail the behavior of various urban subsystems, such as transportation, public facilities, infrastructure, etc.

9.4   Integrating GIS and remote sensing into vulnerability analysis

The rest of this chapter is devoted to illustrating how GIS and remote sensing can be integrated to translate the conceptual framework presented in Figure 9.1 into an applied model for place-based vulnerability analysis. The idea of context particularity implies locational variations in the outcome of vulnerability analysis as a consequence of spatial (and temporal) variations in underlying factors. These locational variations prompt the need for a spatially explicit model of vulnerability analysis. A model is said to be spatially explicit if the inputs and outputs of this model vary according to spatial location (Goodchild and Janelle, 2004). The value of using GIS and remote sensing in translating the proposed conceptual framework into an applied model for urban vulnerability analysis arises directly from the capabilities of these technologies in supporting spatial analysis and decision making, and the generation of place-based knowledge.

Based on the earlier discussion in this chapter, it can be argued that the extent to which GIS and remote sensing technologies are effectively used in the context of vulnerability analysis depends on the ability to balance two competing demands (Rashed and Weeks 2003). The first demand is offering a replicable way for researchers as well as planners and decision makers undertaking local risk reduction efforts to generate concrete profiles of vulnerable communities and to monitor changes in these profiles over time. The second is being able to bring together divergent perspectives and epistemological positions on urban vulnerability in order to test related theories and hypotheses, thus establishing links between place-based and universal levels of knowledge about vulnerability. Such links can ultimately improve our understanding of the interrelations among various contextual factors and global pressures that produce vulnerability patterns.

To meet these demands, Rashed (2006) suggests the following design criteria for integrative GIS and remote sensing place-based vulnerability analysis:

1. Emphasize the use of geospatial resources, i.e. software tools, remotely sensed images, GIS data layers, census data, etc., that are generally available to planners and decision-makers in any reasonably medium-sized urban area.

2. Recognize the divergent perspectives on urban vulnerability.

3. Be multihazards-based.

4. Incorporate policy and more explicit planning components.

5. Generate quantitative parameters that allow for the comparison of differential vulnerability within the focal system.

6. Involve a spatiotemporal modelling engine for urban dynamics that will allow us to collect evidence to support or reject alternative hypotheses concerning the causal linkages between vulnerability, and the social and physical characteristics of urban places, as well as the effects of planning policies.

Building on the above-listed criteria, Rashed (2006) proposed a procedure for place-based vulnerability analysis using GIS and remote sensing. In the following sections, we review this model of urban vulnerability analysis and then report on the findings of a case study that represents an initial attempt to test the applicability of the proposed procedure.

9.5   A GIS–remote sensing place-based model for urban vulnerability analysis

The framework in Figure 9.1 illustrates the degree of complexity involved in vulnerability analysis and draws attention to the value of a place-based analysis in the production of context-derived knowledge of urban vulnerability. Regardless of the spatial scale, the conception of place as a socio-ecological system entails the presence of causal linkages among an array of factors that potentially affect the vulnerability of the coupled human–environment system in a place (Turner *et al.*, 2003). Accordingly, the integrated GIS–remote sensing procedure of place-based vulnerability analysis shown in Figure 9.2 is centred on a dynamic causal model that adopts a systems-thinking approach to explain how vulnerability patterns arise from adverse interactions between and among the components of the socio-ecological system under consideration (Rashed 2006).

Causal models can be orientated in one of two ways: starting with a set of causes and examining their consequences, or starting with a set of consequences and moving down to their causes. The model shown in Figure 9.2 uses the latter path, through a distinctly spatial induction approach to vulnerability analysis. Inductive reasoning acknowledges the particularity of urban places and the need for generating place-based knowledge of vulnerability without assuming any *a priori* hypotheses. Spatial induction means that the problem of vulnerability can be conceptualized as a spatial search problem through which a particular geographic place or region is first screened for evidence of vulnerability. This is done by examining the range of potential losses that may be caused by hazards in an urban place and working back to a measure of the vulnerability of that place. The derived measure of vulnerability

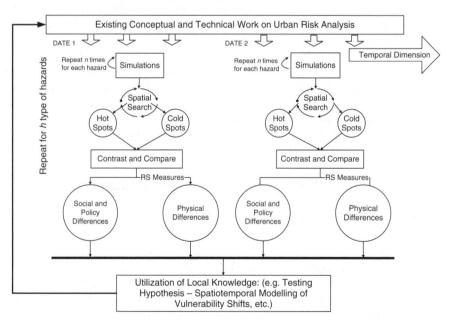

Figure 9.2 Technical framework for the integrative GIS–remote sensing model for place-based urban vulnerability analysis. Adapted from Rashed, T. (2006). In Compagna, M. (ed.), *GIS for Sustainable Development: Bringing Geographic Information Science into Practice towards Sustainability*. Taylor and Francis (CRC Press): New York, NY, USA, 287, 309

is then utilized as an instrument to learn about the range of local factors influencing vulnerability, which might be hidden to the observer or seem quite remote from the hazardous event. This local-generated knowledge can ultimately help devise effective and sustainable risk reduction policies.

To implement the idea of spatial screening, the model proposes the utilization of current advances in geospatial techniques to simulate actual and hypothetical disaster experiences of single or multiple hazards in a particular region. Each simulation will show how potential damages or losses (risks) from a simulated hazard are distributed across the region, assuming that risk = hazard × vulnerability, when several simulations are run using a single set of data pertaining to an urban region at a given point of time (i.e. the particularities of an urban area, and hence vulnerabilities, are controlled for). Variations in simulation results then become a function of the type, location and magnitude of the hazard being simulated. Finding the most vulnerable areas (hot spots of vulnerability) within the urban region then becomes a matter of: (a) ranking urban areas based on the severity of losses calculated from each simulation and (b) searching the region for those areas that maintain relatively high ranks across all the simulation scenarios. These areas are deemed the most vulnerable because maintaining a high rank across different scenarios implies that an area is likely to experience significant losses regardless of the hazard type, originating source or magnitude. Hence, the losses in that place can directly be attributed to its vulnerability. Once areas with high levels of vulnerability are located (the hot spots), spatiotemporal comparisons to areas with lower levels of vulnerability (the cold spots) can be conducted to identify differences and commonalities in their social, physical and political characteristics. As shown in Figure 9.2, the process may be repeated using other datasets that describe the status of the urban region at other points of time. The results can then be utilized to improve our understanding of the relative importance of the various factors influencing vulnerability over space and time, and to dig deeper into the underlying processes amplifying or diminishing vulnerability.

9.6   An illustrative example of model application

To illustrate the utility of the model, we present in this section a first application in a pilot case study from Los Angeles County, California. Due to the exploratory nature of this case study, we have limited our investigation to a single context (Los Angeles County), a single hazard (earthquakes), a single date (1990) and a single question, relating to the links among differential physical and social vulnerabilities to urban earthquakes and urban environmental conditions, as measured from satellite remote sensing. The purpose of the case study is to give a practical example of carrying out place-based vulnerability using GIS and remote sensing technologies. Hence, a full discussion of the technical details encountered in the implementation of this model is beyond the scope of this chapter. We refer interested readers to

Rashed and Weeks (2003) and Rashed *et al.* (2003), in which extensive discussions of the technical developments that have contributed to the present model can be found, especially those related to the simulation of hazards, the identification of vulnerability hot and cold spots, and the quantification of urban morphology through spectral mixture analysis of remotely sensed imagery. In this chapter we will only touch briefly upon the technical issues deemed necessary for demonstrating the utility of the model and for the interpretation of its results.

9.6.1   Study area

The diverse social and physical character of Los Angeles County makes it an ideal study site for testing the capability of using GIS and remote sensing in generating context-specific knowledge of the relative importance of social and physical variables contributing to the overall vulnerability profile of urban communities in this region. Los Angeles County is one of the most populous and ethnically diverse places in the USA (Gordon and Richardson, 1999). Segregation patterns of ethnicity and socio-economic classes in Los Angeles, accompanied by successive waves of economic restructuring and population expansion, have been reflected in the built environment and the physical structure of urban form within the region (Rubin, 1977; Allen and Turner, 1997; Modarres, 1998). For example, Li (1998), comparing areas in Los Angeles dominated by population groups from China and Indochina vs. those dominated by groups from Taiwan and Hong Kong, showed that even the micro-divisions within the same ethnicity have their geographical expression in the spatial differentiation of the region's urban landscape.

The study area has witnessed several earthquake events in the past century. The most recent was an M6.7 earthquake which originated near Northridge on 17 January 1994, in which 57 people were killed, 9000 were injured and damage exceeded \$25 billion (SSC, 1995). The Northridge earthquake has raised many doubts with regard to levels of vulnerability in a modern urban environment generally designed for seismic resistance (Bolin and Stanford, 1998). Therefore, formulating an understanding of the linkages among social and physical vulnerability patterns to earthquake hazards in Los Angeles County can ultimately aid in the formation of policies in anticipation of the problems accompanying urbanization processes and demographic shifts in this dynamic region.

9.6.2   Data

The unit of analysis (focal system) utilized in this case study was the census tract. In this case study, we investigated a total of 1608 census tracts covering approximately $3220\,km^2$ of the entire urbanized area of Los Angeles County. Most of the spatial and aspatial data utilized in the analysis were obtained from the inventory datasets available from the US Federal Emergency Management Agency (FEMA) and built into HAZUS, the software we used for simulating damage loss from earthquakes (FEMA-NIBS, 1999). Data included inventories of building square footage and value, population characteristics from the 1990 census, costs of building repair, and certain basic economic data. Data for transportation and utility lifelines were also included, as well as several layers for faults, geological conditions, and the locations of the epicentres of past earthquakes. In addition, we utilized other population datasets from the US Census Bureau, and digital maps for soil and slope instability and liquefaction potential.

The satellite data utilized in the remote sensing analysis included a subset (3113 lines × 4801 samples) from a Landsat TM image acquired on 3 September 1990 (path 41, row 36). The acquisition date of this image corresponds reasonably well to the 1990 US Census (taken in April 1990). In addition to the multispectral image, a set of 1.0 m spatial resolution aerial photos were used to aid in the validation of the results.

9.6.3 Identifying vulnerability hot spots

Identification of vulnerability hot spots in Los Angeles was accomplished through an empirical model developed by Rashed and Weeks (2003) for the analysis of urban vulnerability to seismic hazards (Figure 9.3). The Rashed–Weeks model combines elements from the techniques of multicriteria evaluation and fuzzy systems analysis (Malczewski, 1999; Jiang and Eastman, 2000) to generate vulnerability scores for urban places. The model was built on top of a robust simulating engine of damage from earthquakes called HAZUS (HAZards in the US) developed by FEMA. HAZUS utilizes methods that have been tested by the State of California Office of Emergency Services and calibrated with data from earthquakes that occurred in sites located within our study area. It also has the capability to generate loss estimates at the census tract level, and this is very important to establish links with social measures of vulnerability derived from census data.

As illustrated in Figure 9.3, there are seven main stages in applying the Rashed–Weeks model of vulnerability analysis. The first stage is the selection of evaluation criteria based on damage estimates to be generated from the simulation. The following criteria have been used as basis of deriving the results presented below (Rashed and Weeks, 2003):

1. Criteria for social risks, including casualties, percentage of households that might seek temporary shelter after a disaster (a proxy for short-term social losses), and total economic cost required for the replacement, reconstruction and recovery of residential buildings (a proxy for long-term social losses).

2. Criteria for physically-induced and engineering risks, including collapse of structures and loss of contents, area of land that might be burned due to induced fire, and amount of debris.

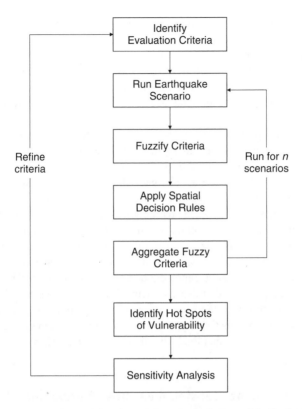

Figure 9.3   Rashed–Weeks model. Adapted from Rashed, T., and J. Weeks (2003) *International Journal of Geographical Information Science.* **17**(6): 547–576

3. Criteria for urban systemic risks which may influence the emergency response and management activities following a disaster, including percentage of loss in functionality for hospitals, fire and police services, power utilities, highways and bridges.

The second stage of the Rashed–Weeks model is the simulation of hazards to explore the combined effects of multiple hazards on a particular region according to multiple scenarios. In the third stage, loss estimates created from a scenario are standardized through a 'fuzzification' process, which recasts values of criteria into statements about set membership using linguistic terms (high, low) (Malczewski, 1999). In the fourth stage, the fuzzified criteria are compared pairwise, using the analytical hierarchy process (AHP) developed by Saaty (1980) in order to generate a set of weights for the evaluation criteria. In the fifth stage, the weighted criteria are aggregated into a one-dimensional array of rules based on a fuzzy additive weighting method. These rules are then used to calculate the membership degree of each census tract in hedged fuzzy sets, which represent the linguistic expressions of the damage states (lower-, medium-, or higher-risk). Stages three to five can be repeated for additional scenarios. In the sixth stage, the 'higher-risk' fuzzy layers produced from all the scenarios are used to locate hot spots of urban vulnerability by identifying those locations that are frequently assigned to higher damage estimates, regardless of the hazard type or source. Finally, in the seventh stage, sensitivity analysis is conducted to determine the effects of simulation parameters on the final output.

The results from applying the Rashed–Weeks model to Los Angeles County based on data from 1990 are presented in Figure 9.4. The maps shown in Figure 9.4A represent the results of the simulation of five earthquake scenarios (four deterministic and one probabilistic). These results were produced by applying the evaluation criteria to obtain a final fuzzy set that represents an index of higher risk in each scenario. Darker areas indicate places with higher damage estimates in the scenario. The map shown in Figure 9.4B represents the distribution of higher-vulnerability values in Los Angeles County derived from the resultant simulation maps of earthquake risks. In this map, darker areas in the figure represent places with higher vulnerability, while brighter areas represent places with lower vulnerability. A visual inspection of the map shows that census tracts with a higher degree of membership in the higher-vulnerability index (i.e. vulnerability hot spots) are clustered in the NW quadrant of Los Angeles County, near the cities of San Fernando and Burbank. As we move away from this quadrant, the degree of membership decreases, and so does vulnerability.

9.6.4 Deriving remote sensing measures of urban morphology in Los Angeles

9.6.4.1 MESMA

The model in Figure 9.2 utilizes remote sensing techniques to understand how the hot and cold spots generated from the simulation physically differ in terms of land cover composition and urban spatial structure. The rationale behind this analysis is that patterns of urban morphology represent the locus of the diversity of engineering, socio-economic and political interactions within urban places. Thus, if differences are found among hot and cold spots of vulnerability in terms of the physical composition and spatial configuration, this could suggest ways in which urban morphology might be manipulated through sustainable policies, to reduce vulnerability to hazards. It could also provide a means to monitor progress toward sustainable hazards mitigation within a giving urban context.

A recurrent theme in several studies in remote sensing has been related to the derivation of summary indicators of the physical components of urban areas. This type of analysis has traditionally been limited due to the spectral heterogeneity of urban features in relation to the spatial resolution of the remote sensors (Weber, 1994), especially true in the context of multispectral images with medium spatial

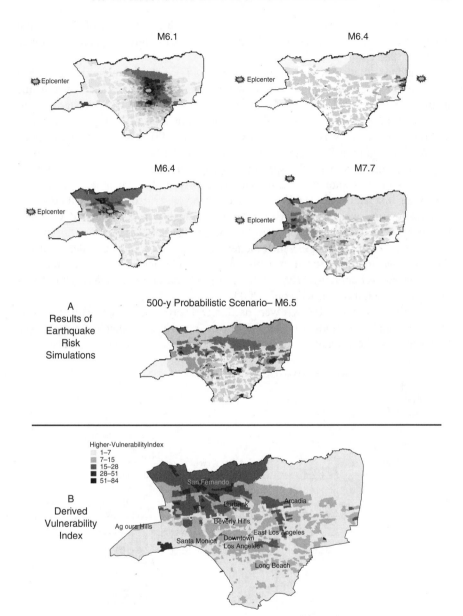

Figure 9.4 Results from applying the Rashed–Weeks model in Los Angeles. Adapted from Rashed, T., and J. Weeks (2003) *International Journal of Geographical Information Science*. **17**(6): 547–576

resolution, such as those provided by Landsat satellites. Because of this spectral heterogeneity, there is a need to deal with a complex mixture of spectral responses (Forster, 1985).

To address the spectral mixing problem and to obtain more representative measures of the composition and structural patterns of urban land cover in the metropolitan area of Los Angeles, the remote sensing analysis task was accomplished in the present case study through the application of multiple endmember spectral mixture analysis (MESMA) (Rashed *et al.*, 2003) and landscape metrics. The MESMA approach, originally developed by Roberts *et al.* (1998), is based on the concept that, although the spectrum in any individual pixel can be modelled with relatively few endmembers, the number and type of endmembers are variable across an image. In this sense, MESMA can be described as a modified linear spectral mixture analysis (SMA) approach, in which many simple SMA models are first calculated for each pixel in the image. The objective is then to choose, for every pixel in the image, which model amongst the candidate models provides the best fit to the pixel spectrum while producing physically reasonable fractions. The procedure of applying MESMA to the 1990 Landsat TM image (Figure 9.5) is described in detail in Rashed *et al.* (2003).

The results from the MESMA were used in two ways to describe spatial variation in the physical conditions between the census tracts in Los Angeles in 1990. The first way was the calculation of an average normalized measure per census tract

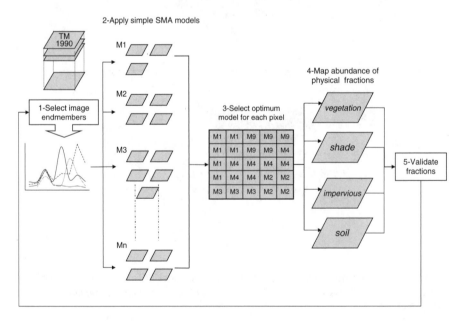

Figure 9.5 An overview of the MESMA approach. Adapted from Rashed, T. et al. (2003) *Photogrammetric Engineering and Remote Sensing* **69**(9): 1011–1020

for each of the four land cover categories of derived MESMA: vegetation, soil, impervious surface and water/shade (Figure 9.5). The normalization was achieved by first summing up the fractional abundance of each category within each census tract, then calculating the ratio of the total fractional abundance to the census tract's area. The product of this process was a normalized value (range 0–100) per census tract for each of the four land cover categories, indicating the average abundance of the land cover within that tract.

The second way of utilizing remote sensing measures in the present study was the derivation of second-order measurements from MESMA fractions that described the configuration (form) of the census tracts in terms of urban land cover. The use of landscape metrics in the analysis of urban landscape patterns is one of the topics that recently received increasing attention in the urban remote sensing community (Geoghegan *et al.*, 1997; Alberti and Waddell, 2000; Parker *et al.*, 2001; Herold *et al.*, 2002, 2003). Landscape metrics are indices developed for categorical map patterns, based on both information theory and fractal geometry (Herold *et al.*, 2002; McGarigal *et al.*, 2002). Categorical map patterns represent data in which the ecosystem property of interest is represented as a mosaic of patches. The definition of patches is imposed according to a phenomenon of interest and only meaningful when referenced to a particular scale (McGarigal *et al.*, 2002). For example, the urban landscape of Los Angeles can be described as a mosaic of census tracts. The census tract in this case can be thought of as a patch that is relatively homogeneous in terms of social and physical conditions. Similarly, at a larger scale, a census tract can be viewed as a mosaic (or landscape) of its own, consisting of smaller patches of land cover classes represented by a collection of pixels.

Unlike the soft classification nature of MESMA results, landscape metrics operate upon a hard or crisp classification assumption. Therefore, before landscape metrics were used in the present study, MESMA fractional images had to be reclassified, such that each pixel within any census tract corresponded to one, and only one, class of land cover. A threshold of 60% was arbitrarily chosen, assuming that when a given land cover class occupies 60% or more of a pixel, then it is possible to say that this pixel generally belongs to that land cover class. When fraction values within a pixel failed to meet this criterion, then a decision role was applied to assign a class to that pixel according to what class the majority of neighbourhood pixels within a 3 × 3 window had.

The next step was to select a subset of landscape metrics to measure the spatial properties of census tracts in Los Angeles. Two types of metrics were used. The first was the class-level metrics, which were applied to zones of land cover types within census tracts (i.e. each zone of land cover category was considered a landscape made of individual pixels or patches). The second type was the census tract-level metrics, which treated each census tract as a landscape made of zones or patches of land cover categories. Tables 9.1 and 9.2 list the subsets of metrics that have been used on either the land cover class or census tract levels.

Table 9.1 Description of landscape metrics applied at the land cover class level within a census tract

Class metrics	
Metric	Property measured
PD (patch density)	Areal composition
LPI (largest patch index)	Areal composition
PAFRAC (perimeter-area fractal dimension)	Shape complexity
PLADJ (percentage of like adjacencies)	Degree of aggregation of land cover class
AI (index of aggregation)	Degree of aggregation of land cover class
IJI (interspersion and juxtaposition index)	Degree of interspersion or intermixing of land cover class
DIVISION	Diversity of land cover class
COHESION	Physical connectedness of the land cover class

Table 9.2 Description of landscape metrics applied at the census tract level

Landscape metrics	
Metric	Property measured
PD – (patch density)	Areal composition
LPI (largest patch index)	Areal composition
PAFRAC – (perimeter-area fractal dimension)	Shape complexity
CONTAG	Overall fragmentation of land cover classes
AI – (index of aggregation)	Degree of aggregation of land cover classes
IJI – (interspersion and juxtaposition index)	Degree of interspersion or intermixing of land cover classes
SIDI – (Simpson's diversity index)	Diversity of land cover classes

9.6.5 Deriving an index of wealth for Los Angeles County

Information on wealth was used in this case study as a proxy for access to resources, which in turn was used as an indication of the distribution of social vulnerability. Although this wealth index is not as sophisticated and comprehensive as other social vulnerability indices proposed in the literature, e.g. that of Cutter *et al.* (2003), we deem it satisfactory for the present study, given its exploratory and illustrative purposes.

To calculate an index of wealth for Los Angeles in 1990, data were used from the US Census Bureau's Survey of Income and Programme Participation (SIPP) to calculate the ratio of wealth to income at each income level by race and by age group. The next step was to use data from the 1990 Public Use Microdata Sample (PUMS) to convert the ratios derived from the SIPP data to the closest income categories that are available in the 1990 census of the study area. The averaged values represented multipliers to be applied to a table that included information on the number of households by income category and race by age for each census tract. Finally, the average household wealth was calculated for each census tract, weighted by the average income, race and age of householders in the census tract. The outcome of this process was a wealth index for Los Angeles County in 1990, which we utilized as an indication of the overall level of access to resources (and hence social vulnerability) in each census tract.

9.6.6 Spatial filtering of variables

Although spatial autocorrelation has long been a concern in geographic literature, it has not yet been routinely addressed in remote sensing applications or in vulnerability analysis (Rindfuss *et al.*, 2004). However, it is well known that data aggregated at particular spatial units, such as census tracts, will be more similar to data for other nearby spatial units than they are to more distant spatial units, because of the bias caused by spatial autocorrelation (Getis and Ord, 1992). Cliff and Ord (1981) identify two general approaches for resolving these problems: (a) filtering spatially autocorrelated data to account for spatial autocorrelation; or (b) modifying statistical models to accommodate spatial autocorrelation (such as spatially autoregressive models).

In the present study we utilized the former approach, following a method of spatial filtering suggested by Getis (1995). Getis' spatial filtering technique involves the extraction of the spatially autocorrelated portion of each of the variables to be input in an ordinary least-squares (OLS) linear regression analysis and then the use of the spatial portion as a separate factor (Getis, 1995; Scott, 1999). By solving the OLS regression model with the extracted filtered and spatial components of the variables, the spatial autocorrelation is removed from the residuals and incorporated into the model to help predict variation in the dependent variable. Summing the absolute values of the statistically significant standardized beta coefficients then allows us to determine the proportion of explained variation that is due to the spatial component, whereas the remainder of the explained variation is accounted for by the filtered (non-spatial) component. The ratio of the square of the beta coefficients for any two independent variables indicates their relative contribution to the prediction of the dependent variable.

9.6.7 Generating place-based knowledge of urban vulnerability in Los Angeles

9.6.7.1 Statistical models

Three statistical models were developed in order to: (a) demonstrate the utility of the model in generating place-based knowledge of the relative importance of the urban morphological social and physical conditions in shaping the spatial patterns of urban vulnerability to earthquakes in Los Angeles County; and (b) compare this place-generated knowledge against conventional wisdom of vulnerability. The first model tested the null hypothesis that the index of wealth (IW), used as a proxy for social vulnerability, was not significantly correlated with the index of higher vulnerability (IV), calculated from the simulation of earthquake risks. The second employed a step-wise OLS regression to examine the extent to which wealth is predicted exclusively by remote sensing measures describing urban physical characteristics. The model employed IW as a dependent variable, and the following independent variables: (a) MESMA fractional measures of vegetation, soil, impervious surface and water/shade normalized by census tract; and (b) landscape metrics calculated as second-order measures of MESMA results (listed in Tables 9.1 and 9.2). The format of this model, after applying the spatial filtering, was as follows:

$$\text{Wealth}(IW) = (\text{normalized MESMA fractions filtered})$$

$$+ (\text{normalized MESMA fractions spatial})$$

$$+ (\text{landscape metrics filtered}) + (\text{landscape metrics spatial}) + \text{error}$$

$$(9.1)$$

The third model was a binary logistic regression model that examined the presence or absence of higher vulnerability (IV) based on values of a set of explanatory variables. Logistic regression was used in this part of the analysis because of the ordinal nature of the fuzzy measure of vulnerability, which allowed for a binary division of the dependent variable into high (1) and low (0) using a threshold value. The explanatory variables used in this third model included the index of wealth (IW), as well as a set of remotely sensed measures that were found to be statistically associated with wealth in the OLS regression model. The general form of this model was:

$$\text{Logit}(Pi) = \log\left[Pi/(1 - Pi)\right] = a + bXi \qquad (9.2)$$

where i represents the binary value of vulnerability, Pi is the conditional probability of Yi given Xi, a is the intercept, b is the vector of slope parameters and Xi is the vector of explanatory variables (wealth and remotely sensed measures).

9.6.7.2 Results of correlation between vulnerability and wealth

Table 9.3 shows Pearson's correlation coefficients between vulnerability and wealth. The table reports a correlation value of 0.11 between vulnerability (IV) and wealth (IW), indicating a low, but nonetheless statistically significant, negative correlation at the 0.01 level, leading us to reject the null hypothesis that wealth, as a proxy for social vulnerability, is not associated with vulnerability values estimated through the simulation of biophysical risks in urban areas. The correlation between the IW and the spatial portion of the IV in Table 9.3 indicates that only the spatial components in the two indexes were significantly correlated, suggesting more evidence for the importance of 'where you are' in the distribution of vulnerability in Los Angeles. While these correlation values were not as high as one may have anticipated, based on what the literature suggests, the significance of such results becomes more apparent if we recall that the IV and IW represent the results of two totally independent methods for measuring vulnerability. Thus, while the negative correlation between wealth and vulnerability found in the model conforms to the universal wisdom, the relatively low correlation value means that the most vulnerable physical elements do not always overlap with the most vulnerable populations within Los Angeles. This finding is important because it is almost identical to what Cutter *et al.* (2000) found from an analysis conducted in Georgetown County, South Carolina, suggesting a pattern that is likely to be common in other urban places in the USA.

Further, some previous studies (e.g. Scott, 1999; Weeks *et al.*, 2000) have suggested the existence of a lag between change in the social environment and the corresponding change that may occur in the physical environment, with the former occurring first. In fact, Scott (1999, pp. 111–112), in the context of her

Table 9.3 Results of correlation analysis between vulnerability and wealth

		"IV"	"IV_sp"	"IV_f"
"IW"	Pearson Correlation	−0.111**	−0.149**	0.016
	Sig. (2-tailed)	.000	.000	.531
"IW_sp"	Pearson Correlation	−0.112**	−0.141**	0.008
	Sig. (2-tailed)	.000	.000	.769
"IW_f"	Pearson Correlation	0.045	−0.068**	0.013
	Sig. (2-tailed)	.073	.007	.601
	N	1561	1561	1561

** Correlation is significant at the 0.01 level (2-tailed)

analysis of accessibility to jobs in Los Angeles, showed that the census tracts at the periphery of Los Angeles County (where higher values of IV exist) were classified as low-income tracts in the 1980 census. However, those tracts themselves became high-income in 1990. This implies a rapid social change that occurred throughout the county in the 1980s that might not yet have been reflected by a physical change in 1990. Thus, one can put forward a proposition that a wealth index based on the 1980 census data might have done a better job than the index used here, which was based on the 1990 census data. It can be suggested, then, that the statistically significant correlation results noted above in fact represent strong evidence of a possible causal linkage between the physical and social conditions of urban places with regard to vulnerability (again conforming to universal wisdom about vulnerability patterns). This is further investigated through the results of the regression models reported in the following subsection.

9.6.7.3 Results of regression models

As a first step in examining whether remotely sensed measures can be used in conjunction with social variables to explain the variation in vulnerability, a step-wise OLS regression model was developed. The model employed the IW as a dependent variable, and a total of 40 independent variables (four Normalized MESMA variables, eight variables resulting from applying landscape metrics at the census tract level, and 28 variables resulting from applying the metrics at the four land cover class levels). The technique of spatial filtering was used to split spatially autocorrelated independent variables into their spatial and non-spatial components.

Table 9.4 Spatially filtered OLS regression for the index of wealth (IW)

Variable	Unstandardized Coefficient	Standardized β	t	Significance of t
Dependent Variable IW				
Impervious_f	−2177.326	−0.0361	−14.763	0.000
IJI_Shade_sp	526.144	0.157	5.777	0.000
Vegetation_f	1748.643	0.184	8.959	0.000
Impervious_sp	−877.699	−0.073	−2.980	0.003
IJI-Shadei_f	206.075	0.075	2.854	0.004
PD_Impervious_f	1532.003	0.394	11.253	0.000
PD_Impervious_sp	1506.867	0.340	10.008	0.000
Vegetation_sp	1475.475	0.055	2.228	0.000
R	0.767			
Adjusted R^2	0.586			
$z(1)$ For residuals	0.89			
N	1561			

Note: see text for an explanation of the variables

The results of the model are shown in Table 9.4, in which only statistically significant predictors (at the 0.05 level) are reported. The R value for this model was 0.767, with an adjusted R^2 of 0.586. An examination of the residuals showed that they were not spatially autocorrelated and exhibited no heteroscedasticity. Also, the results of the co-linearity diagnostic indicated that the independent variables had scored low (< 9) in the condition index. The results show that four of 40 variables utilized emerged as statistically significant predictors of the index of wealth. Among these, two were Normalized MESMA measures (vegetation and impervious surface) and two were derived from landscape metrics applied at the land cover class level within census tracts (PD_Imp and IJI-shd). Considering the absolute values of the statistically significant standardized beta coefficients, we can determine that MESMA measures have accounted for about 26% of the explained variation in the wealth, most of which was related to variation in vegetation. The measures derived from landscape metrics accounted for about 74%. Further, the spatial component in all variables accounted for about 52% of the explained variation in the wealth, while the filtered component accounted for the remaining 48%.

The results in Table 9.4 indicate that the most important predictors of the wealth index were the spatial and non-spatial components of PD_impervious, a landscape metric measure that describes the density of patches within the impervious land cover class in a census tract. The results show that although the density of impervious surface in census tracts is indicative of higher wealth, the abundance of impervious surface fractions derived from MESMA is negatively associated with wealth. This interesting finding highlights the value of applying landscape metrics to MESMA measures to reveal certain physical patterns within an urban place that may not otherwise be shown if one is only relying on the measurement of the physical composition in that place. Table 9.4 also lists vegetation as a strong predictor of wealth, with higher vegetation abundance associated with the more affluent census tracts – a finding that has been reported repeatedly in other urban settings (e.g. Ryznar, 1998; Rashed et al., 2001; Small, 2001).

Finally, results in Table 9.4 indicate that the IJI_shade, another landscape metric applied at the land cover class level, has emerged as a significant predictor of higher wealth. IJI measures the degree of the intermixing of patches within a land cover class. A lower IJI value indicates that patches belonging to a land cover class within a census tract are more aggregated and less fragmented. The results in Table 9.4 suggest that wealth increases (and social vulnerability decreases) with the increase of fragmentation in the shade within a census tract. Since shade has been used in the analysis as a proxy for building heights, one can conclude that tracts with low-rise buildings, e.g. single-family housing, would be characterized with higher IJI values. On the other hand, tracts with high-rise building will possess lower IJI values, and in Los Angeles these areas are likely to score lower on the wealth index, as in the case of downtown Los Angeles. The second regression model utilized was a binary logistic model that used the index of vulnerability (IV) as a dependent variable, and wealth and the remotely sensed measures emerged as statistically significant

Table 9.5 Logistic regression for the index of vulnerability (IV)

Variable	β	*Wald*	Significance	EXP(β)
Dependent Variable IV				
Impervious	0.1390	0.9342	0.3338	1.1491
Vegetation	0.6273	21.1980	0.000	1.8725
IJI_Shade	0.3634	5.8804	0.0164	1.4838
PD_Impervious	0.6987	19.6991	0.000	2.0112
Wealth 1	−0.0723	0.3239	0.5692	0.9303
Wealth 2	0.6018	28.5415	0.0000	1.8253
Wealth 3	0.3628	11.5632	0.0007	1.4451
Wealth 4	−0.2658	5.6609	0.0180	0.7666
Overall percent correct	63.36%			
Chi Square	15.3524		0.0317	
Nagelkerke R^2	0.102			
N	1561			

Note: see text for an explanation of the variables

predictors of the wealth index in the OLS regression model. The results of the model are shown in Table 9.5. The threshold used to determine the binary values of the IV was based on the mean value of the index. Those values that were above the mean were assigned to 1, indicating higher vulnerability, and those values that were equal to or less than the mean were assigned to 0, indicating lower vulnerability. The model was also tested using other thresholds and the results were generally consistent with those listed in Table 9.5. The overall correct prediction of the model was about 63%, with $\chi^2 = 15.34$ at a 0.05 level of significance.

The results in Table 9.5 show that three out of the four remotely sensed variables utilized emerged as statistically significant predictors of higher vulnerability. The strongest among these was again the landscape metric-based measure, PD_impervious, the higher values of which were shown to increase the odds of being highly vulnerable by a factor of 2.01, holding all other variables constant. On the other hand, as expected, being in the higher wealth category (wealth 4) reduces the odds (by a factor of 0.77) of being in the highly vulnerable category. This suggests that the wealth (social) effect is independent of the remotely sensed (physical) effect, and that both need to be taken into account if we are to understand the vulnerability of place.

9.6.8 To what extent do model results conform to universal knowledge of vulnerability?

The purpose of this case study has been to provide an applied example of the utility of an integrative GIS–remote sensing model for place-based vulnerability analysis. Generated knowledge of vulnerability was used to fulfil two objectives: first, to explore the basic hypothesis that social vulnerability is manifested through aspects from the physical environment in urban places within Los Angeles; and second, to examine the proposition that remote sensing can provide us a quantitative means to describe and assess aspects related to urban spatial structure that influence vulnerability in that region.

To address the first objective, we examined the correlation between the wealth index and vulnerability. The results showed a statistically significant negative correlation between the two indexes, although not high enough to conclude that the wealth can be taken as a sole indicator of vulnerability. Given the apparent difference between the spatial distributions of values in the two indexes, an obvious question arises: how do these results conform to theories of vulnerability found in the literature? The answer to this question can be discussed in light of the relationship between access to resources and vulnerability. This relationship was previously examined by researchers in the context of disasters in developing countries (e.g. Wisner, 1993; Blaikie et al., 1994). In these studies, access to resources was traditionally measured by the level of poverty determined by income (as opposed to the concept of wealth utilized here). In developing countries, spatial and physical aspects of vulnerability tend to be much more pronounced because the poor are often forced to live and work persistently in hazardous areas (Hewitt, 1997). In contrast, socially and economically marginalized populations in the USA do not necessarily live in areas at greatest risk of natural hazards (Bolin and Stanford, 1999). Indeed, the wealthy people may even choose to live in physically hazardous settings, such as earthquake-prone hillsides in California (Davis, 1998). Therefore, vulnerability in this case has little to do with systematic differences between the rich and poor in terms of their exposure to the earthquake, a finding confirmed above in the model results.

Additionally, the general literature on vulnerability draws a distinction between two patterns of vulnerability: *persistent* (or *chronic*) vulnerability and *situational* vulnerability (Bolin and Stanford, 1998). Persistent vulnerability connects to social forces that produce economically, ethnically and culturally marginalized groups. Situational vulnerability, on the other hand, occurs when some population groups (including wealthy and financially secured ones) become increasingly at risk in the face of calamity. This might happen due to a combination of circumstances related to their jobs, choice of housing, etc., but does not necessarily need to be related to social or demographic factors. That is, in situational vulnerability, a household has the option to choose not to live in a hazardous place. In persistent vulnerability, the social factor is much more noticeable, while the physical aspect of vulnerability is implicit. Situational vulnerability is quite the opposite case, in which the physical aspect of vulnerability becomes more apparent and the social aspect becomes implicit. It is our contention that these patterns of persistent and situational vulnerabilities were represented respectively by the index of wealth (IW) and the index of vulnerability (IV) produced by the simulation of physical damage resulting from earthquake scenarios. The mismatch of the spatial distribution between the two indexes implies some missing information related either to social vulnerability (in the case of the IW) or physical vulnerability (in the case of IV).

The second objective fulfilled by the knowledge generated in this case study is related to the utility of remote sensing for providing measures that can be used as surrogates for social vulnerability. The results of the OLS model showed that the remotely sensed variables accounted for about 57% of the explained variation in the IW. The results of the logistical regression showed that the remotely sensed variables emerged as significant predictors of the IV. The moral of these results is that remote sensing data can be used to derive information about the physical composition and spatial structure of the built environment in an urban place. This information reflects aspects of the social environment that will be manifested in the demography and culture of people. The built environment, represented by the arrangement of land cover classes, then interacts with the socio-economic environment (measured, at a minimum, by income, race and ethnicity) to produce the urban environment. The urban environment then creates a difference in people's vulnerability by influencing the volume and intensity of social interaction, which in turn has implications for the opportunities that exist for different social groups to access resources.

There is no doubt that a small number of statistical models based on one unique urban area in a developed country cannot be taken as a foundation upon which to build a grand theory of vulnerability to disasters, or to explain how vulnerability is reflected in the urban spatial structure. However, the results of these models are still sufficient to draw attention to the utility of place-based vulnerability analysis using GIS and remote sensing in obtaining information that addresses core issues of the social sciences such as social vulnerability.

9.7   Conclusions

The disaster caused by Hurricane Katrina in the USA in 2005, and the subsequent course of events that shaped the disaster in affected cities along the US Gulf Coast, revealed a striking example of physical and social vulnerabilities in 'western' cities in their worst-case scenario. The disaster has strongly challenged, or at least shown the need for revisiting, some popular views that are frequently portrayed in the literature in either an implicit or explicit manner, for example, the idea that 'inhabitants of less developed countries [are] more likely to die from hazards than those in more developed ones' (Bankoff, 2004, p. 29), or the emphasis on development as an exclusive means to reducing risks (UNDP, 2004). These kinds of broad generalizations with regard to vulnerabilities and risks could be misleading, because there is no place or group of people that can be thought of as entirely safe, neither is there a magic single solution for reducing urban risks. Rather, vulnerability exists in each urban society across the globe but is manifested in different forms. These could be underdevelopment in one society, lack of education and technology in a second, poor urban governance in a third, failure to translate knowledge into action in a fourth, or a combination of two or more of these and other forms.

As Hewitt (1997, p. 143) underscores, 'vulnerability analysis is essentially about the human ecology of risks'. Ecological factors that are embedded in the landscape of an urban place contribute in different ways to the overall vulnerability pattern of that place. These ecological factors represent, in varying degrees, the context-altering forces that drastically affect people's resilience and ability to cope with and recover from losses. They also provide a means to uncover and understand differential vulnerability within and between urban places. Yet, because these ecological factors are variable and do not hold a constant relationship among themselves, no two urban places are likely to be found that are identical in their vulnerabilities. As a result, it is difficult to develop a broadly applicable action plan that can be followed to diagnose vulnerability and reduce disaster impacts in every single place in the world. Therefore, as we have strongly argued throughout this chapter, revealing context particularities and being decisive for context-sensitive mitigation policies are essential goals of urban vulnerability analysis.

In this chapter, we have capitalized upon the idea of particularity and proposed a conceptual framework for analysing vulnerability across nested scales of urban socio-ecological systems. We have shown how GIS and remote sensing can be integrated to translate this framework into a replicable model for place-based vulnerability analysis. We showed through a wall-to-wall exercise an initial attempt to apply this model to analyse urban vulnerability to earthquake hazards in Los Angeles County, California. Despite the limited scope of the analysis that was carried out, the results of the model call attention to some key considerations that underline the potential of our GIS–remote sensing model for place-based urban vulnerability assessment. The first is that stratification of potential disaster impacts is strongly influenced by a range of contextual conditions, both societal and organizational, which may not be directly related to the geophysical mechanisms of the triggering of hazardous events. The second is the central role of urban dynamics modelling as a means to better understand differential vulnerabilities in cities. The third consideration is that, although vulnerability is largely a reflection of conditions created and modified by human actions, one cannot discard the fact that knowledge of the geophysical properties of natural hazards is essential to understand how dangers arise at the interface of society and natural conditions. Finally, reducing losses from hazardous events is not a problem that can be solved in isolation through a traditional urban planning model. Rather, it requires an understanding of the magnitude of shock that a given urban system is prepared to absorb while remaining capable of operating, and of the means to build management models that take into account the long-term impacts of mitigation efforts on current and future generations. Future developments and applications of our model will need to be expanded in order to ensure that these considerations are equally balanced.

Our model depends upon an integration of GIS and remote sensing. Thus far, the main stream of GIS and remote sensing integration discussions is devoted to addressing practical details. Technical issues, such as whether and how the coupling of GIS and remote sensing should be loosely or tightly implemented, common interface design, building of hybrid remote sensing-GIS databases, data sharing and interoperability, etc., have been, and continue to be, central to most of the discussions (Ehlers, 1990; Mesev, 1999; Chen et al. 2000; Longley and Mesev, 2001; Chen, 2002; Longley, 2002). Few researchers (e.g. Mesev, 1997; Rindfuss and Stern, 1998; Rindfuss et al., 2004) moved beyond the narrow technical detail to larger methodological issues involved in the integration of the technologies under the umbrella of GIS, for example, problems of spatial autocorrelation, spatial–temporal mismatch, classification compatibility, etc., but attempts made in this regard remain technical in tone and very generic, easy to acknowledge but difficult to resolve.

There is no doubt that technical issues are central to GIS and remote sensing integration. Naturally, we have encountered lots of technical details and methodological challenges in the course of developing and applying the place-based vulnerability analysis model, some of which we were able to resolve, while others remain an avenue for future developments. We have also learned the importance of seeking guidance from the subject matter (i.e. urban vulnerability in Los Angeles) to inform the development and integration of the technologies and the selection of solution options. That is, we have learned how the fields of vulnerability and hazards can help inform the selection, development and integration of GIS and remote sensing techniques as much as we learned about the tools GIS and remote sensing can offer to vulnerability analysis. For example, the use of a simulation approach in deriving different scenarios of damage resembles to a greater extent the way in which disaster managers traditionally utilize past disaster experiences as instruments to learn about the adverse consequences of hazardous events in cities, and to infer the underlying factors that need to be addressed to promote the level of safety in the community. We used this very basic idea to develop algorithms that can screen a multitude of disaster scenarios back into a measure of vulnerability of the place. Likewise, our use of MESMA and landscape metrics in quantifying the physical dimension of urban morphology in Los Angeles was inspired both by the characteristics of the physical settings of our study site and by discussions in the vulnerability literature about how the characteristics of the urban spatial structure (e.g. open spaces, land use/land cover, transportation layout) influence the function of the city in the immediate aftermath and during the recovery from disaster impacts (Hewitt, 1997; Menoni et al., 2000). This use of subject matter in guiding the development of the GIS–remote sensing integrative model exemplifies the way in which universal wisdom of vulnerability can be used to guide the investigation into the particularities of place discussed earlier in this chapter.

To this end, we suggest that the integration argument in the ongoing GIS–remote sensing literature needs to be extended further beyond its current technical and methodological focus to include the subject matter or phenomenon under consideration; how its underlying dynamics vary over space, and how established theories in such fields as economic, political and social sciences can be used to inform remote sensing–GIS integration. Earlier in the chapter, we argued that urban places can be used as an analytical basis for urban vulnerability analysis. In the conclusions of this chapter, we again argue for urban places, or space in general, but this time to be used as a basis for a wider concept of GIS–remote sensing integration, not only in terms of data but also in terms of the development of functions, algorithms and models that acknowledge the unique challenges each place brings to GIS–remote sensing analysis and can ultimately provide a basis for contextually aware decision making.

Acknowledgements

The research presented in this paper was partially supported by a grant from the National Science Foundation (BCS-0117863). The case study reported in this chapter was presented at the 3rd International Conference of Urban Remote Sensing, Regensburg, Germany, June 2003.

References

Alberti, M. and Waddell, P. (2000) An integrated urban development and ecological simulation model. *Integrated Assessment* **3**(1): 1215–1227.

Allen, J. P. and Turner E. (1997) *The Ethnic Quilt: Population Diversity in Southern California*. The Centre for Geographical Studies, California State University: Northridge, CA, USA.

Anderson, R. E., Carter, I. and Lowe G. R. (1999) *Human Behavior in the Social Environment: A Social Systems Approach*, 5th edn. Aldine De Gruyter: New York, NY, USA.

Andrews, H. F. (1985) The ecology of risk and the geography of intervention: from research to practice for the health and well-being of urban children. *Annals of the Association of American Geographers* **74**, 370–382.

Bankoff, G. (2004). The historical geography of disaster: 'vulnerability' and 'local knowledge' in Western discourse. In Bankoff G. *et al.* (eds), *Mapping Vulnerability: Disasters, Development and People*. Earthscan: Sterling, VA, USA, 25–36.

Blaikie, P., Cannon, T. Davis, I. and Wisner, B. (1994) *At Risk: Natural Hazards, People's Vulnerability, and Disasters*. Routledge: New York, NY, USA.

Bolin, R. and L. Stanford (1998) *The Northridge Earthquake: Vulnerability and Disaster*. Routledge: New York, NY, USA.

Bolin, R. and L. Stanford. (1999). Constructing Vulnerability in the First World: The Northridge Earthquake in Southern California, 1994. In Oliver-Smith, A, and Hoffman, S. M. (eds), *The Angry Earth: Disaster in Anthropological Perspective*. Routledge: New York, NY, USA, 89–112.

Burton, I., Kates, R. W. and White, G. F. (1978) *The Environment as Hazard*. Oxford University Press: New York, NY, USA.

Cardona, O. D. (2004). The need for rethinking the concepts of vulnerability and risk from a holistic perspective: a necessary review and criticism for effective risk management. In Bankoff, G. *et al.* (eds), *Mapping Vulnerability: Disasters, Development and People.* Earthscan: Sterling, VA, USA, 37–51.

Chen, K. (2002) An approach to linking remotely sensed data and areal census data. *International Journal of Remote Sensing* **23**(1): 37–48.

Chen S, Zheng, S. and Xie, C. (2000) Remote sensing and GIS for urban growth in China. *Photogrammetric Engineering and Remote Sensing* **66**(10): 593–598.

Cliff A. D. and Ord, J. K. (1981) *Spatial Processes: Models and Applications.* Pion: London, UK.

Cova, T. J. (1999). GIS in emergency management. In Longley, P. A.*et al.* (eds), *Geographical Information Systems.* Wiley: New York, NY, USA, 845–858.

Cutter, S. L. (1996) Vulnerability to environmental hazards. *Progress in Human Geography* **20**(4): 529–539.

Cutter, S. L. (2001) A research agenda for vulnerability science and environmental hazards. *IHDP Update: Newsletter for the International Human Dimensions Programme on Global Environmental Change* **2**(1): 8–9.

Cutter, S. L. (2003a) GI science, disasters, and emergency management. *Transactions in GIS* **7**(4): 439–445.

Cutter, S. L. (2003b) The vulnerability of science and the science of vulnerability. *Annals of the Association of American Geographers* **93**(1): 1–12.

Cutter, S. L., Boruff, B. J. and Shirley, W. L. (2003) Social vulnerability to environmental hazards. *Social Science Quarterly* **84**(1): 242–261.

Cutter, S. L., Mitchell, J. T. and Scott, M. S. (2000) Revealing the vulnerability of places: a case study of Georgetown County, South Carolina. *Annals of the Association of American Geographers* **90**(4): 713–737.

Davis, M. (1998) *The Ecology of Fear.* Metropolitan: New York, NY, USA.

Dow, K. (1992) Exploring differences in our common feature(s): the meaning of vulnerability to global environmental change. *Geoforum* **23**(3): 417–436.

Ehlers, M. (1990) Remote sensing and geographic information systems: towards integrated spatial information processing. *IEEE Transactions on Geoscience and Remote Sensing* **28**(4): 763–766.

FEMA–NIBS (Federal Emergency Management Agency and Institute of Building Sciences). (1999) *HAZUS: User's Manual and Technical Manuals*, Vols 1–3. FEMA–NIBS: Washington, DC, USA.

Fitzpatrick, K. and LaGory, M. (2000) *Unhealthy Places: The Ecology of Risk in the Urban Landscape.* Routledge, New York, NY, USA.

Forster, B. C. (1985) An examination of some problems and solutions in monitoring urban areas from satellite platforms. *International Journal of Remote Sensing* **6**(1): 139–151.

Geoghegan, J., Wainger, L. A. and Bockstael, N. E. (1997) Spatial landscape indices in a hedonic framework: an ecological economics analysis using GIS. *Ecological Economics* **23**(3): 251–264.

Getis, A. (1995). Spatial filtering in a regression framework: examples using data on urban crime, regional inequality and government expenditure. In Anselin, L. and Florax, R. (eds), *New Directions in Spatial Econometrics.* Springer-Verlag: Berlin, Germany.

Getis, A. and Ord, J. K. (1992) The analysis of spatial association by use of distance statistics. *Geographical Analysis* **24**; 189–206.

Goodchild, M. F. and Janelle, D. G. (2004). Thinking spatially in the social sciences. In Goodchild, M. F. and Janelle, D. G. (eds), *Spatially Integrated Social Science*. Oxford University Press: New York, NY, USA, 3–22.

Gordon, P. and Richardson, H. W. (1999) Review essay: Los Angeles, City of Angels? No, City of Angles. *Urban Studies* **3**: 575–591.

Herold, M., Goldstein, N. and Clarke, K. (2003) The spatiotemporal form of urban growth: measurement, analysis and modelling. *Remote Sensing of Environment* **86**: 286–302.

Herold, M., Scepan, J. and Clarke, K. C. (2002) The use of remote sensing and landscape metrics to describe structures and changes in urban land uses. *Environment and Planning A* **34**: 1443–1458.

Hewitt, K. (1997) *Regions of Risk: a Geographical Introduction to Disasters*. Longman: Harlow, UK.

Hewitt, K. and Burton, I. (1971) *The Hazardousness of a Place: a Regional Ecology of Damaging Events*. Published for the University of Toronto Department of Geography by University of Toronto Press: Toronto, Canada.

ICSU. (2002) *Science and Technology for Sustainable Development*. ICSU Series on Science for Sustainable Development, No. 9. International Council for Science: Paris, France.

Jiang, H. and Eastman, J. R. (2000) Application of fuzzy measures in multi-criteria evaluation in GIS. *International Journal of Geographic Information Science* **14**(2): 173–184.

Kates, R. W. (1971) Natural hazards in human ecological perspective: hypotheses and models. *Economic Geography* **47**(3): 438–451.

Li, W. (1998) Anatomy of a new ethnic settlement: the Chinese ethnoburb in Los Angeles. *Urban Studies* **35**(3): 479–501.

Liverman, D. M. (1990). Vulnerability to global environmental change. In Kasperson, R. E. *et al.* (eds), *Understanding Global Environmental Change: The Contributions of Risk Analysis and Management*. Clark University: Worcester, MA, USA, 27–44.

Longley, P. A. (2002) Geographical information systems: will developments in urban remote sensing and GIS lead to 'better' urban geography? *Progress in Human Geography* **26**(2): 231–239.

Longley, P. A. and Mesev, V. (2001) Measuring urban morphology using remotely-sensed imagery. In Donnay, J.-P. *et al.* (eds), *Remote Sensing and Urban Analysis*. Taylor and Francis: London, UK, 163–183.

Malczewski, J. (1999) *GIS and Multicriteria Decision Analysis*. Wiley: New York, NY, USA.

McGarigal, K., Ene, E. and Holmes, C. (2002) *FRAGSTATS: Spatial Pattern Analysis Program for Quantifying Landscape Structure*. University of Massachusetts, Amherst, MA, USA: http://www.umass.edu/landeco/research/fragstats/fragstats.html

Menoni, S. *et al.* (2000) Measuring the seismic vulnerability of strategic public facilities: response of the health care system. *Disaster Prevention and Management* **9**(1): 29–38.

Mesev, V. (1997) Remote sensing of urban systems: hierarchical integration with GIS. *Computers, Environment and Urban Systems* **21**(3/4): 175–187.

Mesev, V. (1999) Editorial: integration issues in GIS and remote sensing. *Computers, Environment and Urban Systems* **23**(1): 1–3.

Mileti, D. S. (1999) *Disasters by Design: a Reassessment of Natural Hazards in the United States*. Joseph Henry Press: Washington, DC, USA.

Mitchell, J. K. (1989). Hazards research. In Gaile, G. L. and Willmott, C. J. (eds), *Geography in America*. Merrill: Columbus, OH, USA, 410–424.

Mitchell, J. K., Devine, N. and Jagger, K. (1989) A contextual model of natural hazards. *Geographical Review* **79**(4): 391–409.

Modarres, A. (1998) Putting Los Angeles in its place. *Cities* **15**(3): 135–147.

Mustafa, D. (2005) The production of an urban hazardscape in Pakistan: modernity, vulnerability, and the range of choice. *Annals of the Association of American Geographers* **95**(3): 566–586.

Parker, D. C., Evans, T. P. and Meretsky, V. (2001) *Measuring Emergent Properties of Agent-Based Land-cover/Land-use Models Using Spatial Metrics*, Vol. 2002. Seventh Annual Conference of the International Society for Computational Economics: http://php.indiana.edu/~dawparke/parker.pdf. Seventh Annual Conference of the International Society for Computational Economics: http://php.indiana.edu/~dawparke/parker.pdf

Radke, J. *et al.* (2000) Application challenges for GIScience: implications for research, education, and policy for risk assessment, emergency preparedness and response. *Journal of the Urban and Regional Information Systems Association* **12**(2): 15–30.

Rashed, T. (2006) Geospatial technologies, vulnerability assessment, and sustainable hazards mitigation in cities. In Campagna, M. (ed.), *GIS for Sustainable Development: Bringing Geographic Information Science into Practice towards Sustainability*. Taylor and Francis (CRC Press): New York, NY, USA, 287, 309.

Rashed, T., and J. Weeks (2003) Assessing Vulnerability to Earthquake Hazards through Spatial Multicriteria Analysis of Urban Areas, *International Journal of Geographical Information Science*. **17**(6): 547–576.

Rashed, T., J. Weeks, M. Gadalla, and A. Hill (2001) Revealing the Anatomy of Cities through Spectral Mixture Analysis of Multispectral Satellite Imagery: A Case Study of the Greater Cairo Region, Egypt, *Geocarto International*. **16**(4): 5–16.

Rashed, T. *et al.* (2003) Measuring the physical composition of urban morphology using multiple endmember spectral mixture models. *Photogrammetric Engineering and Remote Sensing* **69**(9): 1011–1020.

Rejeski, D. (1993). GIS and risk: a three-culture problem. In Goodchild, M. F.*et al.* (eds), *Environmental Modelling with GIS*. Oxford University Press: Oxford, UK, 318–331.

Rindfuss, R. R. and Stern, C. (1998). Linking remote sensing and social science: the need and the challenges. In Liverman, D. M. (ed.), *People and Pixels: Linking Remote Sensing and Social Science*. National Academy Press: Washington, DC, USA, 1–27.

Rindfuss, R. R. *et al.* (2004) Developing a science of land change: challenges and methodological issues. *Proceedings of the National Academy of Sciences of the USA* **101**(39): 13976–13981.

Roberts, D. A. *et al.* (1998) Mapping Chaparral in the Santa Monica mountains using multiple endmember spectral mixture model. *Remote Sensing of Environment* **65**: 267–279.

Rubin, B. (1977) A chronology of architecture in Los Angeles. *Annals of the Association of American Geographers* **67**(4): 521–537.

Ryznar, R. M. (1998). Urban Vegetation and Social Change: an Analysis using Remote Sensing and Census Data. PhD Dissertation, University of Michigan, Ann Arbor, MI, USA.

Saaty, T. L. (1980) *The Analytic Hierarchy Process*. McGraw-Hill: New York, NY, USA.

Scheurer, T. (1994) *Foundations of Computing: System Development with Set Theory and Logic*. Addison-Wesley: Cambridge, MA, USA.

Scott, L. M. (1999). The Accessible City: Employment Opportunities in Time and Space. Doctoral Dissertation, San Diego State University/University of California at Santa Barbara, CA, USA.

Small, C. (2001) Estimation of urban vegetation abundance by spectral mixture analysis. *International Journal of Remote Sensing* **22**(7): 1305–1334.

SSC (Seismic Safety Commission of the State of California). (1995) *Northridge Earthquake: Turning Loss to Gain*. SSC: Sacramento, CA, USA.

Tobin, G. A. and Montz, B. E. (2004) Natural hazards and technology: vulnerability, risk, and community response in hazardous environments. In Brunn, S. D. *et al.* (eds), *Geography and Technology*. Kluwer Academic: Dordrecht, The Netherlands, 547–570.

Turner, B. L. II *et al.* (2003) Science and technology for sustainable development: a framework for vulnerability analysis in sustainability science. *Proceedings of the National Academy of Sciences of the USA* **100**(14): 8074–8079.

UNDP (2004) *Reducing Disaster Risk: A Challenge for Development*. United Nations Development Programme, Bureau for Crisis Prevention and Recovery: New York, NY, USA.

Weber, C. (1994) Per-zone classification of urban land cover for urban population estimation. In Foody, G. M. and Curran, P. J. (eds), *Environmental Remote Sensing from Regional to Global Scales*. Wiley: Chichester, UK, 142–148.

Weeks, J. R. *et al.* (2000) Spatial variability in fertility in Menoufia, Egypt, assessed through the application of remote-sensing and GIS technologies. *Environment and Planning A* **32**(4): 695–714.

White, G. F. and Haas, J. E. (1975) *Assessment of Research on Natural Hazards*. MIT Press: Cambridge, MA, USA.

Wisner, B. (1993) Disaster vulnerability: scale, power, and daily life. *GeoJournal* **30**(2): 127–140.

Wisner, B., Blaikie, P., Cannon, T. and Davis, I. (2004) *At Risk: Natural Hazards, People's Vulnerability, and Disasters*, 2nd edn. Routledge: London, UK.

10

Using GIS and remote sensing for ecological mapping and monitoring

Jennifer A. Miller* and John Rogan†

*Department of Geography and the Environment, University of Texas at Austin, TX, USA

† Graduate School of Geography, Clark University, Worcester, MA, USA

10.1 Introduction

The ability to map and monitor ecological phenomena over large spatial extents has become a focus of renewed research in the context of increasing awareness of human activities and environmental change (Busby, 2002; McDermid *et al.*, 2005; Liu and Taylor, 2002). Human activities substantially impact most of the terrestrial biosphere, currently at rates and spatial extents far greater than in any other period in human history (Kerr and Ostrovsky, 2003). Numerous organizations, disciplines and initiatives have formed in the last 15 years in response to the myriad challenges to sustainable resource management and ecological protection, e.g. the International Association of Landscape Ecology, the NASA Land Cover/Land Use Change Program. These interdisciplinary and integrative initiatives agree that scientifically sound and sustainable resource management requires ecological data of variable spatial and temporal characteristics to provide the scientific understanding required to measure, model, maintain and/or restore landscapes at multiple scales (EPA, 1998; Wiens *et al.*, 2002). Research efforts in support of sustainable ecosystem management have focused on characterizing ecosystem condition and change, exploring the effects of different management schemes, and understanding how natural and anthropogenic processes affect ecosystem functioning (EPA, 1998).

Integration of GIS and Remote Sensing   Edited by Victor Mesev

© 2007 John Wiley & Sons, Ltd.

Solutions to these problems require spatially explicit, timely, ecological data, often combined with statistical models in a geographic information system (GIS).

Current research illustrates how ecological problems ranging from biodiversity loss to land-use change have benefited greatly from advances in geospatial technologies such as GIS and remote sensing, both in the provision of data and access to spatial data analysis tools. The integration of GIS and remote sensing for ecological mapping and monitoring, while addressed in earlier research (Stoms and Estes, 1993; Franklin, 1995; Goodchild, 1994), has become even more important as these data and technologies continue to evolve, and as ecological issues become more critical. The key motivations for integrating GIS and remote sensing for ecological research and management are:

1. The acceptance of the landscape context and scale for sustainable ecosystem management (Liu and Taylor, 2002).

2. The importance of retrospective and prospective monitoring for conservation (Urban, 2002; Turner et al., 2003).

3. Increased familiarity with GIS and remote sensing data and methods within resource management agencies (Jennings, 2000).

4. Improved geospatial data quality and availability (at reduced cost) (Rogan and Chen, 2004).

5. Reported advantages of using different types of geospatial data (from both GIS and remote sensing) for mapping and monitoring applications (Rogan and Miller, 2006; Zimmermann et al., 2007).

Although the benefits of integrating GIS and remote sensing data for more effective ecological mapping and monitoring are many, the time, money and expertise required to take full advantage of the technology can be initially daunting. The information used for scientifically valid ecological mapping and monitoring needs to be frequently updated, sufficiently detailed and spatially continuous. Ecological inventories have historically been conducted through field survey – a time-consuming and expensive endeavour, particularly when study sites are large and/or remote, and when long-term monitoring is a concern to resource managers (Rogan and Chen, 2004). This field work paradigm has implicitly affected both the typical study area size and the spatial scale of observations associated with ecological research. Field data are also typically collected based on some purposive sampling scheme, in which information on a specific ecological attribute (e.g. species abundance, timber inventory) is of primary interest, and therefore may not be appropriate for describing other attributes of subsequent interest (e.g. productivity, fuel loadings, habitat suitability). Lack of familiarity and background knowledge, equipment cost and complexity of data-processing methods are often cited as factors that prevent even wider use of remote sensing approaches by ecologists as well as by practitioners in other disciplines (McDermid et al., 2005; Treitz and Rogan, 2004).

The availability of ecological datasets, collected through remote sensing (e.g. land-cover, NDVI) or derived within a GIS (e.g. topographic moisture index, incoming solar radiation) at local to global scales, has revolutionized the way ecological research is conducted (Cohen and Justice, 1999; Rushton *et al.*, 2004). GIS enhances the ability to derive information from remotely sensed data, and remotely sensed data can describe actual environmental conditions for expedient updating of GIS databases. The synoptic perspective, temporal frequency and repeatability of remotely sensed measurements have been invaluable for detecting and monitoring change (Rogan *et al.*, 2003).

In a recent review of remote sensing applications in ecological research, Kerr and Ostrovsky (2003) identified three main application focus areas: land-cover classification, integrated ecosystem measurements and multitemporal change detection. This chapter examines the ways in which remotely sensed data have been integrated with GIS data and modelling approaches in the context of these three areas. We focus on species distribution models (SDM)[1] and biodiversity mapping/modelling as particular cases of ecological mapping, we summarize the GIS and remotely sensed environmental data that are most commonly used in these applications, and we include a case study that integrates GIS and remote sensing for ecological monitoring (land-cover change mapping). Although the traditional inconsistency in spatial scale between remotely sensed data (indirect) and ecological field observations (direct) has been a major obstacle to more extensive integration of remote sensing in ecological research, access to increasingly fine spatial resolution data has resulted in great progress in this area (Turner *et al.*, 2003; Aplin, 2005; Kerr and Ostrovsky, 2003). Further, advances in remote sensing theory, data and technology over the past 35 years have led to general and robust methods of large area data collection that, for many ecological attributes, can provide more reliable estimates than field methods (Davis and Roberts, 2000). The integration of GIS and remotely sensed data and techniques can greatly facilitate all steps of data collection, compilation, analysis and visualization. The potentially synergistic benefits of integrating GIS and remotely sensed data with statistical methods are still being explored and identified.

The three main sources of information used in ecological mapping and monitoring applications are shown in Table 10.1. Field observation provides the most detailed and fine-scale information, although the spatial coverage is not continuous. Field data are also expensive and time-consuming to collect, and many of the observations are relatively subjective or suited for a narrow purpose. GIS data can provide continuous spatial coverage (usually through interpolation methods), albeit at coarser

---

[1] We follow the convention of Guisan and Thuiller (2005) in using 'species' to refer to both plants and animals, as animal species habitat suitability is directly related to plant species habitat suitability. Although we use the term 'species' here, these models can also be used to predict species assemblages (Ferrier *et al.*, 2002; Franklin, 1995).

Table 10.1 Characteristics of field, GIS and RS data sources used for ecological mapping and monitoring

Data collection	Benefits	Limitations
Field observation	Fine spatial scale Detailed information Direct observation	Limited temporal extent Incomplete spatial coverage Expensive Subjective
GIS	Associated with potential distributions Can be used for species-level mapping Can be used to derive direct and resource gradients	Limited spatial resolution Unknown accuracy Cannot be frequently updated Grid cell values usually result from interpolation Indirect gradients most readily available
RS	Associated with actual distributions Allows data collection in remote areas Synoptic perspective Systematic measurement for every pixel; Complete spatial coverage Enable larger study areas Multitemporal; high temporal resolution Cost-effective for large extents	Atmospheric obstructions possible Expensive for fine spatial scales Less detailed information Processing methods intimidating (to untrained users) Usually represent indirect or functional gradients
GIS and RS (integrated)	Data can be upscaled More consistent and objective databases Can provide updated environmental data Data are readily available Direct, resource, indirect, and functional gradients can be combined	Compounding of quantitative–positional errors Lack of automated methods to aid integration Paucity of raster GIS data with fine spatial resolution (to match RS data)

spatial resolution and lower or unknown positional accuracy. GIS data, particularly digital elevation models (DEMs), have been used to derive complex environmental variables that are more ecologically relevant (e.g. topographic moisture index, potential solar radiation). Remote sensing facilitates data collection in difficult- or impossible-to-reach areas and provides an important synoptic and multitemporal perspective. Remote sensing also systematically provides a value for each pixel and spatially continuous coverage. However, fine spatial resolution imagery consistent with the scale of field observation is expensive, some amount of processing is required, and atmospheric obstructions can be problematic. The integration of GIS and remotely sensed data improves upon many of their individual limitations.

10.2 Integration of GIS and remote sensing in ecological research

Rogan and Miller (2006) summarized four ways in which GIS and remote sensing data can be integrated: (a) GIS can be used to manage multiple data types; (b) GIS analysis and processing methods can be used for manipulation and analysis of remotely sensed data (e.g. neighbourhood or reclassification operations); (c) remotely sensed data can be manipulated to derive GIS data; and (d) GIS data can be used to guide image analysis to extract more complete and accurate information from spectral data.

Remotely sensed data and techniques have been widely available since the early 1970s but the most common ecological application for which they are combined with GIS is still mapping land cover (Stoms and Estes, 1993). The coarsened scale of these land-cover maps, as well as the potential for circular reasoning when using them to model plant species distributions, renders them unsuitable for most SDM studies (Zimmermann et al., 2007). The continuous properties of image spectral values and vegetation indices have rarely been used as predictor variables in SDM, although both show great potential (Frescino et al., 2001; Osborne et al., 2001; Suárez-Seoane et al., 2002; Zimmermann et al., 2007). GIS has been a mainstay of SDM through the derivation of and analysis with bioclimatic factors associated with species distributions (see Franklin, 1995; Guisan and Zimmermann, 2000) but other factors, such as competition and disturbance, may be more appropriately described by remotely sensed data. Despite the synergistic potential of combining GIS and spectral data, remote sensing is rarely used directly in ecological mapping studies (but see section 10.5).

10.3 GIS data used in ecological applications

We use 'GIS data' here to describe non-spectral digital environmental data, as they are stored, manipulated and typically derived in a GIS. These data are derived either by interpolating field or station observations to a continuous surface (e.g. temperature) or by calculating new surfaces from existing spatially continuous data (e.g. slope from a DEM). The availability of digital data that represent increasingly complex environmental characteristics provides the basis for SDM (for reviews, see Franklin, 1995; Guisan and Zimmermann, 2000; Guisan and Thuiller, 2005). Climatic and topographical variables are the most widely used predictors in SDM, as they describe broad-scale physiological tolerances related to water and temperature, and site energy and moisture availability associated with micro-climates, respectively (Franklin, 1995). These environmental variables are reviewed below, along with a summary of gradient analysis, which provides the conceptual framework for describing the way in which environmental gradients influence species distributions.

10.3.1 Gradient analysis

SDM evolved from research methods that used gradient analysis to explore how plant species composition and distribution change along environmental gradients (Whittaker, 1973; Kessell, 1979; Franklin, 1995). The range of environmental conditions a species is physiologically able to tolerate defines its fundamental niche, which is analogous to its potential distribution. Due to other factors, such as competition or disturbance, a species typically only occupies a subset of its fundamental niche, which is termed its 'realized niche', or actual distribution. SDM involves the quantification of the species–environmental gradient relationship, the result of which is a species habitat[2] distribution map. The nature of the gradient variable determines how robust the resulting model is likely to be. Austin (1980) describes three types of environmental gradients:

1. *Direct gradients* are those in which the environmental variable has a direct physiological effect on species growth but is not consumed (e.g. temperature).
2. *Resource gradients* have a direct physiological effect on growth, and are actually used or consumed (e.g. water, nutrients).
3. *Indirect gradients* have no direct physiological effect and are likely the result of a location-specific correlation with one or more direct gradients (e.g. correlation between temperature and elevation makes elevation an indirect gradient).

Data describing indirect gradients are usually more readily available in digital format, or easily measured in the field, and are often highly correlated with observed species patterns (Guisan and Zimmermann, 2000). However, these relationships are location- and gradient-specific, usually describing a combination of other direct or resource gradients, and models in which they are used are generally not appropriate for extrapolation beyond the area in which the data were collected (Franklin, 1995; Austin and Gaywood, 1994). Variables that represent direct or resource gradients are the most suitable for extrapolating across space and time, although Franklin (1995)

---

[2] Recent articles have discussed the ambiguity in the use of the word 'habitat' (de Leeuw *et al.*, 2002; McDermid *et al.*, 2005). Here we consider habitat to refer to the type of environment (as measured by a suite of environmental factors) in which an organism normally occurs.

points out the incongruity of using empirical data on actual species distribution (realized niche) to map potential species distribution (fundamental niche).

A fourth type of gradient added by Müller (1998), '*functional*' gradients, is associated with species response to direct and resource gradients. Many of the spectral-derived variables used in ecological applications of remote sensing, such as productivity, biomass, and leaf area index, can be described as functional gradients. Models that include functional gradients allow for accurate depiction of *actual* landscape composition, structure and function (Rollins *et al.*, 2004) and may be particularly useful for modelling disturbed landscapes (Lees and Ritman, 1991; Frescino *et al.*, 2001). Functional gradients should be considered location-specific in the same sense as indirect gradients, although Zimmermann *et al.* (2007) discuss an exception involving the use of spectral-derived variables that describe vegetation structure as direct gradients for modelling bird species distribution.

The distinction between mapping actual vs. potential vegetation species has been discussed previously (Woodcock *et al.*, 2002; Franklin, 1995). Maps of potential species distribution are used in models to investigate effects of climate change on vegetation distribution, and as the first step in parameterizing dynamic vegetation models (Guisan and Zimmermann, 2000). However, actual species distribution maps may be more appropriate for certain applications, such as resource management or biodiversity measurement. Figure 10.1 illustrates the typical flow of information and data types used when GIS and remotely sensed data are integrated for ecological mapping and monitoring. A SDM based solely on GIS data, such as bioclimatic variables, makes the often untenable assumption that species distributions are at equilibrium with their environment (Guisan and Zimmermann, 2000). The resulting potential habitat distribution maps, particularly for species whose distributions have been modified by anthropogenic effects, are likely to be over-predicted, as they describe the fundamental rather than the realized niche (Thuiller *et al.*, 2004). Coarse-scale predictor variables, such as climate, are typically correlated with potential distributions, while finer-scale topographic variables are associated with actual distributions (Thuiller *et al.*, 2003; Franklin, 1995). Guisan and Thuiller (2005) further divide the hierarchical effects into (from global to local scale) limiting climatic factors, dispersal factors, disturbance factors and resource factors. Additional information on biotic and abiotic factors that reduce a species' fundamental niche must be included in the SDM to produce a map of actual species distribution. Functional gradients that describe actual ecosystem characteristics, such as normalized difference vegetation index (NDVI), can be used as predictor variables to produce actual species distribution.

While many direct and resource gradients important to species distributions are still unknown, immeasurable or difficult to describe across a landscape (Whittaker, 1973), the ability to derive increasingly complex (and more ecologically relevant) environmental variables using GIS and remote sensing has great potential. Accurate and timely spatial information describing actual ecological characteristics is essential for predicting future conditions, and remotely sensed data

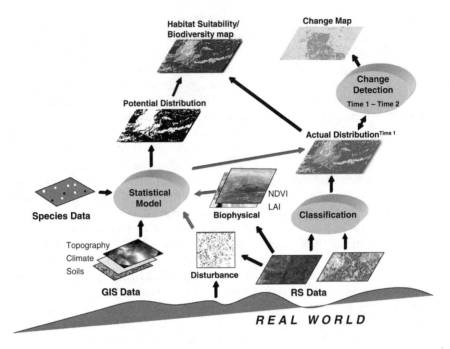

Figure 10.1 Typical information flow for integrating GIS and RS in ecological mapping and monitoring applications

are particularly useful for this purpose. The next generation of gradient models will incorporate direct, resource and functional gradients to more accurately map ecological characteristics (Franklin, 1995; Austin and Heyligers, 1989).

10.3.2 Climate

Climate has been linked with vegetation distribution from at least the early 1800s, when von Humboldt wrote about the relationship between latitude and vegetation type (as cited in Jongman *et al.*, 1995). Early global plant distribution maps based on climatic factors alone were surprisingly accurate (e.g. Holdridge, 1947, 1967) and the relationship between climate and vegetation remains very important in broad-scale vegetation modelling. In addition to providing the fundamental relationship on which static vegetation maps are based, an understanding of the complex feedback relationship between vegetation and climate variability is necessary to parameterize dynamic vegetation models used to study global biogeochemical cycles.

Climate data are particularly important in SDMs used for predicting consequences of global warming on plant and animal distributions. Many plants and animals are limited in their distribution by temperature extremes. Teixeira and Arntzen (2002) observed that slight changes in climate will have a particularly important effect on ectothermal (cold-blooded) animals, as temperature directly affects most of their physiological processes, as well as other important environmental conditions (air humidity, soil moisture, and vegetation composition). Thuiller *et al.* (2003) found that temperature and precipitation extremes were effective surrogates for bioclimatic factors with more direct physiological roles (e.g. evapotranspiration) in limiting the distribution of Mediterranean vegetation in Spain. While climate variables, such as temperature and precipitation, tend to have broad-scale influence on species distributions, many finer-scale bioclimatic indices, such as potential solar radiation, mean relative humidity and potential evapotranspiration, can be derived using other GIS data (e.g. elevation, aspect) and may be more directly related to species distributions (Franklin, 1998; Cairns, 2001; Leathwick, 1998; Meentemeyer *et al.*, 2001).

Thuiller *et al.* (2004) found that climate was the most important driver of species distributions (plants, mammals, birds, reptiles and amphibians) in Europe. They observed that, at the relatively coarse scale of their study (50 km), land-cover provided largely redundant information relative to climate. However, using finer-scale land-cover data (1 km) to predict plant species presence in Britain, Pearson *et al.* (2004) found that availability of suitable land-cover was more important than availability of suitable climate. Cumming (2002) suggested that the seasonality of precipitation may be as important as its magnitude, particularly for animal distributions. Climate can influence animal distribution both directly (e.g. bird migration in winter) and indirectly (e.g. land cover and food availability) (Venier *et al.*, 2004). Although their importance in determining species distributions is well established, digital climate surfaces have generally been produced by interpolating ground station data and are of limited availability and quality for many areas (Parra *et al.*, 2004). In a recent study, Suárez-Seoane *et al.* (2004) explored the use of climate variables derived from METEOSAT in a model of bird distribution in Spain. They found that these 5 km resolution data showed great potential as an alternative to interpolated climate surfaces, particularly in areas where meteorological stations were sparse.

10.3.3 Topography

While bioclimatic factors, such as water availability, temperature and insolation, are the main drivers of species distributions, topographic variation modifies their influences, resulting in increased spatial heterogeneity associated with microclimatic effects. Although simple topographic variables that represent indirect gradients, such as elevation, slope and aspect, should have less influence when used along with direct and resource gradients in models, they are often empirically important, as they tend to be derived with higher accuracy (Rollins *et al.*, 2004; Guisan and Zimmermann, 2000). One general assumption has been that as the processing steps involved in deriving a topographic variable increase, so too does its susceptibility to error (Guisan and Zimmermann, 2000), although Van Niel *et al.* (2004) note that this is not always the case. In a study that simulated error propagation in the derivation of topographic variables, they found that in some cases more complex variables, such as net solar radiation, were less affected by error than comparatively simple variables, such as slope and aspect (Van Niel *et al.*, 2004). In a similar study, Holmes *et al.* (2000) found that topographic variables derived by compounding values from a large number of other grid cells were more error-prone, and that while global error estimates may be low, local error could be quite high.

Derived topographical variables used in SDM include potential solar radiation, landscape position, slope curvature and topographic moisture index (for reviews, see Florinsky, 1998; Franklin, 1995; Moore *et al.*, 1991). Potential solar radiation can be used to simulate a direct gradient that describes potential evapotranspiration and soil moisture. Landscape position, the location of a grid cell relative to surrounding grid cells (upslope or downslope) is related to a combination of soil properties, specifically depth, texture and potential moisture (Franklin, 1995). Slope curvature and topographic moisture are both related to the water availability of a site (Moore *et al.*, 1991). In a study relating mammalian species richness to environmental variables, Tognelli and Kelt (2004) used elevation range for each quadrat as a proxy variable to represent habitat heterogeneity. White *et al.* (2005) found that topographic variation influenced the way in which vegetation responded to interannual climatic fluctuations (e.g. ENSO), although it is often not explicitly addressed in climate models. Topography also influences the onset, rate, pattern and duration of disturbance intensity and severity (Rogan and Miller, 2006).

Other GIS variables, such as geology and soil type, can also be used to represent moisture and nutrient availability, although usually at a coarser scale due to their measurement level. GIS variables, such as 'distance to _____ (roads, water, edge, etc.)', can represent proximity to disturbance or important resources (Osborne *et al.*, 2001). Landscape metrics, such as pattern, structure and heterogeneity, can also be quantified and used as predictor variables (McGarigal and McComb, 1995; Gottschalk *et al.*, 2005). Habitat suitability maps for other species can also be used to represent potential competition or predation, or can be used to stratify sampling schemes for more rare species (Edwards *et al.*, 2005) .

10.4   Remotely sensed data for ecological applications

Over the past 35 years, remotely sensed data have steadily become an invaluable information source for ecological characterization and survey (see recent reviews by Gong and Xu, 2003; Coppin *et al.*, 2004; Aplin, 2005; Turner *et al.*, 2003; Kerr and Ostrovsky, 2003; Rogan and Chen, 2004). This is primarily due to the effectiveness of air-borne and space-borne remote sensing platforms and sensors that facilitate observation of biophysical attributes over extensive areas at multiple spatial, spectral and temporal scales (Stow, 1995; Jensen, 2000). Landsat (30m), long considered the 'workhorse' of terrestrial remote sensing, has provided the longest-running time series of remotely sensed data at scales appropriate for regional studies. Thus, there is a perceived space–time mismatch between the information that ecologists want and what remotely sensed data can provide. Nonetheless, field data (ground reference) are often limited for regional-global scale investigations because they are rarely as widespread or timely as remotely sensed data (Pettorelli *et al.*, 2005).

The last 8 years have witnessed a proliferation of satellite platforms with a large number of sensors (e.g. Terra and ENVISAT) and improved spatial resolutions (e.g. IKONOS-2 and Quickbird data have pixels that cover an area of $16\,m^2$ or less) that can also serve the needs of timely and cost-effective resource management (Franklin, 2001). Our focus is on 'passive' remote sensing systems, although we acknowledge the benefits of 'active' systems (see Davis and Roberts, 2000; Kasischke *et al.*, 1997).

Remotely sensed data are used primarily in ecological research to characterize land cover, describe habitat structure and derive measurements of biophysical properties. The ability of remotely sensed variables to act as surrogates for important ecological characteristics (e.g. biodiversity, productivity) is a function of the closeness of the relationship between the measured radiation and the environmental variable of interest. 'State' variables are those that can be described directly by the measure of electromagnetic radiation, such as leaf area index and biomass (Curran *et al.*, 1998; Curran, 2001). However, it is most often environmental variables that are indirectly related to the actual radiation measure, such as biodiversity and productivity, that are of interest to ecologists. There are a wide variety of options to choose from to exploit known relationships between values of optical/microwave data and the biophysical properties of ecological entities (for reviews, see Franklin, 2001; Davis and Roberts, 2000). The following section presents an overview of commonly used remote sensing data enhancements and key ecological targets typically investigated.

10.4.1   Spectral enhancements

Considering passive optical data only, numerous spectral transformation methods have been developed to concentrate and accentuate the biophysical signal from the surface into an enhanced spectral 'feature' (Roberts *et al.*, 1998). Spectral vegetation indices (VIs) have been used since the late 1960s, with continued evolution of new types of VI and uses. Advances in technology for both spectral sensing platforms and analytical techniques have led to a wide range of applications for VI, ranging from evapotranspiration estimates to forest structure quantification (Wulder, 1998).

The normalized difference vegetation index (NDVI) has become the most extensively used VI in ecological remote sensing. NDVI is a dimensionless spectro-radiometric measurement derived from optical remotely sensed data that is correlated to micro- and macro-level characteristics of plants (for review, see Pettorelli *et al.*, 2005). It was initially developed as a measure of green leaf biomass (Tucker, 1979) but has also been used effectively in mapping other vegetation attributes, such as percentage cover, stem density, stand health, etc.

Since its introduction almost 30 years ago, the Kauth Thomas (KT or Tasselled Cap) image transformation has proved to be versatile in ecological remote sensing applications (see Kauth and Thomas, 1976; Crist and Ciccone, 1984). This transformation involves the statistical rotation of multispectral data space into a set of physically meaningful VIs that describe scene brightness, greenness and wetness. Brightness has positive loadings in all reflectance bands and corresponds to overall scene brightness, or albedo (Crist and Ciccone, 1984). Greenness, like many other correlates of vegetation amount (e.g. NDVI), is a contrast between the visible bands (especially Landsat TM band 3) and the near-infrared (Landsat TM band 4). Wetness presents a contrast of the visible and near-infrared bands (weak positive loadings) with the mid-infrared bands (strong negative loadings). Cohen *et al.* (1995) found that wetness was least sensitive to topographic variation, and therefore more powerful for predicting forest structural attributes. KT variables have been used in a variety of ecological applications, such as forest canopy mapping and selective harvest detection (Cohen and Fiorella, 1998).

10.4.2 Land cover

Spectral data have been used most often to derive some variation of a map of land-cover type (vegetation, biotype) or quality (biomass, NPP), from which habitat distribution (Osborne *et al.*, 2001; Suárez-Seoane *et al.*, 2002; Venier *et al.*, 2004), measures of abundance (Luoto *et al.*, 2002a) or biodiversity (Luoto *et al.*, 2002b; Debinski *et al.*, 1999; Waser *et al.*, 2004; Tognelli and Kelt, 2004) are classified (see section 10.5). Land-cover mapping determines the current composition and distribution of landscape attributes, and this is subsequently used as the basis for assessing future change. Regions (Homer *et al.*, 1997, Huang *et al.*, 2003), nations (Cihlar *et al.*, 2003), continents (Stone *et al.*, 1994) and the globe (Hansen *et al.*, 2000; Belward *et al.*, 1999) have been mapped at various spatial resolutions with a range of remotely sensed data inputs. Satellite imagery efficiently provides information about vast areas and is, therefore, a useful tool for land-cover mapping across large extents. Neither aerial photography nor field data can provide equal amounts of information as efficiently (Franklin *et al.*, 2003).

Digital image classification is a common approach for predicting the categorical class membership (e.g. forest type) of an observation (pixel), based on spectral band response values. Multitemporal and multisensor spectral measurements, along with GIS data, can be used to add explanatory variables to an analysis that may help discriminate categories of interest (e.g. forest vs. grassland). Variations in the structural attributes of the forest stand may have a greater effect on the reflectance characteristics than tree species composition. Therefore, other mapped environmental variables associated with, or controlling, forest vegetation distributions, such as those related to terrain (digital elevation models), geology, soils, climate or land use, can be combined with image data in the classification process in various ways to aid forest type discrimination.

10.4.3 Habitat structure

A number of studies beginning in the 1950s have empirically and theoretically explored the relationship between pixel-level reflected absorbed and transmitted radiation and habitat structure characteristics. Most studies that estimate vegetation structural and biophysical parameters from remotely sensed data have used empirical methods to relate spectral data and various image derivatives to vegetation characteristics. If these parameters are strongly correlated with remotely sensed data, they can be used to predict those biophysical attributes over large extents (Woodcock et al., 2001).

Structural variables allow ecologists to better discriminate forest habitat types for conservation planning purposes. Based on the framework of Diamond (1988), habitat quality represents the diversity of resources as defined by habitat diversity and structural complexity (i.e. in both vegetative structure and types) (Stoms and Estes, 1993). Using remotely sensed data, forests are often characterized in terms of inventory parameters, which provide detailed data on the location, arrangement, distribution and pattern of forest resources (Wulder, 1998). Generally, inventory parameters such as vegetation type, canopy cover and canopy height are linked to habitat structure which has been recommended for assessing wildlife resources and planning conservation efforts (Roberts and Davis, 2000).

Remote sensing approaches have involved the characterization of canopy or crown cover (also described as crown closure – the vertical projection of vegetation onto the ground when viewed from above) with a good deal of success (Davis and Roberts, 2000). Authors have attributed the success of these studies to the dominance of canopy cover in the radiation ecology of their site environments. A variety of approaches (using passive and active remote sensing instruments) have been developed to map canopy cover because this variable has been a primary motivator of environmental/biophysical remote sensing research (Franklin, 2001). Canopy height is an important hybrid variable in forest biodiversity studies (Davis and Roberts, 2000). Several researchers have reported relationships between canopy height and spectral measurements (Franklin et al., 1986; Cohen and Spies, 1992; Danson and Curran, 1993; Jakubauskas and Price, 1997). Variability caused by topographic variation is often problematic in most landscapes, resulting in imprecise representations of vegetation structural properties (Franklin *et al.*, 2003). Topographic effects may be minimized either by stratifying the study area and forest type to zero-slope or by including a slope variable in regression models. Stand density is defined as the number of individual trees per unit area. Stand density is limited in its applicability to forest ecosystem studies, where measures such as cover, height and volume are more commonly required for habitat characterization and prediction (Davis and Roberts, 2000). Density, however, may be of greater importance in semi-arid and arid regions (more heterogeneous cover), where vegetation density and cover are more highly variable across the landscape (Franklin and Turner, 1992).

Remotely sensed data have been employed in the measurement of many structural variables other than those discussed above, such as stand volume (Ardo, 1992, Oza *et al.*, 1996), basal area (Franklin *et al.*, 1986; Danson and Curran, 1993) and fire fuels estimation (Stow, 1995; Cosentino *et al.*, 1981). Several researchers have adopted comprehensive approaches to measuring forest stand structural variables and have either attempted to determine the total stand structural factor contribution to stand spectral response, or have evaluated the total stand structural information contained within various spectral vegetation indices (Danson and Curran, 1993; Jakubauskas and Price, 1997; Steininger, 2000). This approach can help ecologists learn more about the contributions of component hybrid variables to overall spectral response, rather than only knowing the relationship between spectral response and a single hybrid variable.

10.4.4 Biophysical processes

An increasing amount of attention is being paid to the representation of variables that represent ecological processes using remotely sensed data. Remotely sensed measures of ecological process and productivity include leaf area index (LAI) and net primary productivity (NPP) (Wulder, 1998). Productivity is seldom measured directly (in the field) but is estimated from associated variables, such as temperature, precipitation, solar insolation, actual and potential evapotranspiration, biomass or leaf area index (Davis and Roberts, 2000).

Leaf area index (LAI) is the standard expression for the leaf area of a plant community, defined as the total leaf area per unit ground cover. LAI is an important biophysical attribute of plants because of its potential as an indirect measure of vegetation canopy energy, gas and water exchanges (Chen and Black, 1992). Maximum LAI has been correlated with mean annual temperature, length of the growing season, mean annual minimum air temperature and water availability (Gholz, 1982; Wulder, 1998). Field-measured LAI measures are strongly correlated to VIs, especially NDVI (Chen and Guilbeault, 1996). Unfortunately, the relationship between LAI and NDVI is frequently non-linear, and can be erroneously lower due to canopy shading in mature forest stands. Stratification of NDVI images by vegetation or land-cover class is therefore often used for robust estimation over regional scales (Wulder, 1998).

Net primary productivity (NPP) is defined as the net flux of carbon from the atmosphere into green plants per unit time. NPP refers to a rate process, i.e. the amount of vegetable matter produced per day, week or year. Estimates of NPP are based on ecological models which require detailed inputs, many of which are feasible only when acquired using remote sensing (Wulder *et al.*, 2004). Numerous studies have shown that NDVI is related to ecosystem function, particularly NPP (Friedl *et al.*, 1994; Ramsey *et al.*, 1995). Running (1990) incorporated surface temperatures derived from advanced very high resolution radiometry (AVHRR) with annually integrated NDVI to provide better estimates of NPP.

10.5 Species distribution models

Species distribution models (SDMs) have long been a staple in resource conservation and management efforts, as well as research on the effects of climate change. The most commonly used medium is a map of plant species distribution, which can subsequently be used to derive maps that show suitable habitat characteristics for particular animals (Scott *et al.*, 2002; Franklin, 1995, Guisan and Zimmermann, 2000; de Leeuw *et al.*, 2002; Woodcock *et al.*, 2002). These SDM rely on the digital availability of important environmental variables that influence plant (and subsequently animal) distributions. The product of SDMs, habitat suitability maps, can be used to show current distributions, identify possibly suitable (potential) habitat currently unoccupied, and predict the probable effects of changing environmental conditions. Guisan and Zimmermann (2000) reviewed the increasingly large variety of statistical methods used to quantify the species–environment relationship and discuss some of the conceptual considerations important in method selection. Although basic statistical methods (e.g. linear regression) are now available as part of many GIS software packages, many of the assumptions they make about data (e.g. independent observations, linear relationships between response and predictors) are violated with biogeographical data. More sophisticated statistical analysis is usually done using dedicated or user-written statistical software (for recent overview, see Guisan and Thuiller, 2005). Austin (2002) observes that the continued use of inappropriate statistical methods for SDMs stems from a long-standing disconnect between the ecological knowledge of statisticians and the statistical abilities of ecologists. Austin (2002) notes that linear relationships are still often used in models that describe species–environment relationships, despite both ecological theory and empirical evidence that refute this (Austin, 1987, 2002; Bio *et al.*, 1998). The use of remotely sensed data and the functional gradients they describe (Müller, 1998) in SDMs increases the need for more exploratory and flexible statistical methods.

Table 10.2 summarizes selected recent studies that have integrated GIS and remotely sensed data to model species distribution for inventory, atlas or biodiver-

Table 10.2 SDM studies and descriptions of data used

Study	Response	GIS data	Spectral data	Model*	Location
Austin et al. (1996)	Bird presence	Topography, number of buildings, road length	Land cover, habitat heterogeneity (Landsat-5 TM)	GLM, discriminant function analysis	Scotland
Cumming (2002)	Tick species	Climate, topography, vegetation type, political regions	NDVI (AVHRR)	GLM	Africa
Dirnböck et al. (2003)	Plant communities	Topography	NDVI, reflectance, texture (CIR photographs)	CCA	Austrian Alps
Franklin et al (2000)	Plant communities	Topography	NDVI, BGW, red (band 3), near-IR (band 4), mid-IR (band 5) (Landsat TM)	CT	Southern California, USA
Frescino et al. (2001)	Forest presence, basal area, cover (%), density	Climate, topography, geology	NDVI (AVHRR), Land cover, red (band 3), near-IR (band 4), mid-IR (band 5) (Landsat-5 TM)	GAM	Utah, USA
Gibson et al. (2004)	Bird presence	Topography, 'distance to coast', 'distance to creek'	Habitat complexity (multi-spectral video-graphic image)	GLM	Australia
Jeganathan et al. (2004)	Bird presence	Relative abundance of bushes, trees	NDVI, reflectance (Landsat-7 ETM⁺)	GLM	India

Luoto et al. (2002)	Plant species richness	Topography, habitat structure, soil moisture	Land cover map (Landsat-5 TM, used with DEM)	GLM	South-western Finland
Osborne et al. (2001)	Bird presence	Topography, proximity to disturbance	NDVI (AVHRR)	GLM	Spain
Parra et al. (2004)	Bird presence	Climate, topography	NDVI (AVHRR)	EE	Ecuadorian Andes
Pearson et al. (2004)	Plant presence	climate	Land cover (Landsat)	ANN	UK
Suárez-Seoane et al. (2002)	Bird presence	Topography, road, town, river distance and density	Principal components of NDVI (AVHRR)	GAM	Spain
Thuiller et al. 2004	Plant, mammal, amphibian, reptile, bird presence	Climate	Land cover (AVHRR)	GAM	Europe
Venier et al. (2004)	Bird presence	Climate	Land cover (Landsat MSS and AVHRR)	GLM	Great Lakes, USA
Zimmermann et al. (2007)	Tree presence	Topography, climate	NDVI, BGW, reflectance (Landsat-7 ETM$^+$)	GLM	Utah, USA

*ANN, artificial neural network; CCA, canonical correspondence analysis; EE, environmental envelope; GAM, generalized additive models; GLM, generalized linear models; CT, classification tree.

sity applications. Climate and topography are the most often used GIS variables, and land cover and NDVI are the most often used spectral variables. Climate and NDVI are quite similar with respect to the biotic and abiotic factors they represent, although NDVI generally has higher spatial resolution and lower temporal resolution.

Venier *et al.* (2004) compared MSS-derived land-cover maps to those derived from AVHRR as predictor variables (along with climate) in a model to predict distribution of bird species. They found that, despite differences in spatial resolution, both MSS and AVHRR land-cover maps produced similar results, and that they could not discriminate between the direct and indirect (e.g. through land cover) effects of climate (Venier *et al.*, 2004). Frescino *et al.* (2001) found that both classified and raw TM data resulted in better predictions of forest structure compared to AVHRR-derived NDVI. They suggested that classified spectral data, such as land cover, can provide more information than raw spectral data because ecological characteristics and neighboring pixels add context during the classification process (Frescino *et al.*, 2001). In models used to predict vegetation distribution in Southern California, Franklin *et al.* (2000) found that topographical variables were useful to discriminate between physiognomically similar (and therefore spectrally similar) chaparral species. NDVI has also been used to characterize habitat type by calculating a maximum value composite (MVC) over a 12 month period (Osborne *et al.*, 2001) or as a surrogate for energy (Cumming, 2002). Principal components analysis of the MVC data has also been used to decompose the time series into a sequence of spatial and temporal components (Suárez-Seoane *et al.*, 2002).

Some studies have investigated the relationship between environmental factors and vegetation indices by using VI as a response variable. In a regression tree model with greenness vegetation index (GVI) as the response variable, Michaelsen *et al.* (1994) found that vegetation type was the most important predictor variable, followed by burning treatment (burned areas experienced greater greenness). The relationship between elevation and GVI was more complex and they observed that, during the growing season, elevation acted as a surrogate for soil properties but that after May the relationship became inverse, possibly suggestive of temperature/moisture differences (Michaelsen *et al.*, 1994). In a model for the western USA, Stoms and Hargrove (2000) found that precipitation, temperature and available soil water capacity were the most important predictors of NDVI in undisturbed areas. Their model predictions deviated from actual NDVI values in areas of urban or agriculture land use (Stoms and Hargrove, 2000).

Landscape pattern and heterogeneity are also important predictors in SDM. In a habitat model for an endangered bird species in India, Jeganathan *et al.* (2004) found that density of bushes and trees, as measured on the ground and remotely, was the most important predictor variable. They compared ground-based bush/tree density information to that derived from satellite imagery and found that models based on the imagery performed better, although they suggest that ground-based surveys may be more directly relevant for habitat management decisions (Jeganathan *et al.*, 2004).

Additional 'spatial' habitat variables can be subsequently derived from a digital map of habitat complexity/characteristics (Coops and Catling, 2002).

Statistical models that combine spatial and spectral data have also been used to produce maps for subsequent analysis, such as maps of fuel loads and fire regimes (Rollins et al., 2004) and to provide input maps for dynamic landscape simulation models (Franklin, 2002). Habitat suitability maps for several species have been combined to indicate possible levels of biological diversity within an area.

10.5.1 Biodiversity mapping

Biological diversity, or biodiversity, refers to variability within species, among species and in ecosystems. The applications discussed here are concerned with variability among species, also referred to as species richness. The ability to measure and monitor biodiversity, particularly critical in the context of environmental change (Nagendra, 2001), requires adequate, updated species inventories, as well as detailed knowledge of species–habitat relationships (Scott et al., 1993). The separate effects of climate change and land cover change on habitat loss worldwide are substantial, however, their interaction could be even more devastating (Travis, 2003).

SDMs have been used to produce habitat suitability maps for several different species, which have been combined and used as a surrogate of biodiversity (Ferrier, 2002). Availability of species data is often the limiting factor for this technique (Ferrier, 2002). Remotely sensed variables have also been used as a surrogate for biodiversity, most often by relating it to NPP, for which NDVI is used as a surrogate (Kerr and Ostrovsky, 2003; Aplin, 2005; Skidmore et al., 2003; Turner et al., 2003).

Skidmore et al. (2003) explain the correlation between NDVI and biodiversity as being a function of the debated diversity–productivity hypothesis in ecology. Increased availability of resources in highly productive ecosystems generally results in a greater number of species in a given area. Empirical evidence suggests that the relationship is actually unimodal, with maximum biodiversity associated with some optimal value of productivity, beyond which biodiversity begins to decrease. Skidmore et al. (2003) summarize possible explanations for this.

While others have used AVHRR NDVI data to confirm a unimodal relationship between species richness of plants and productivity (Oindo and Skidmore, 2002), Seto et al. (2004) did not find a quadratic model to be a significantly better fit than a linear model for bird and butterfly species richness in the Western USA. While they observed a positive correlation between bird and butterfly species richness and NDVI mean, maximum and standard deviation, they were unable to identify any functional relationships. However, they note that the relationships vary, based on scale, location and taxonomic group (Seto et al., 2004). Linear relationships may be observed empirically if the range of the species studied extends beyond the study area sampled, as Austin (2002) noted for species response to environmental gradients.

Seto *et al.* (2004) suggest that using NDVI as a proxy for species richness improves upon more subjective and time-consuming classification approaches, which require some *a priori* (and often subjective) correlation between species richness and land cover. As a result, resource managers and ecologists may evaluate more quickly and efficiently the potential biodiversity impacts of alternative management and conservation strategies. Luoto *et al.* (2002b) determined that environmental variables derived from a combination of satellite imagery (Landsat TM) and elevation data provided a reasonable surrogate for plant species diversity. Cumming (2002) developed generalized linear models to predict tick distribution in Africa and found that monthly climate variables (temperature extremes, precipitation) added more information than monthly mean NDVI values, while mean annual NDVI performed better than annual climate variables. He suggested that this was related to the temporal lag in the response of NDVI to climatic changes, rendering NDVI a less sensitive variable at finer temporal resolutions (Cumming, 2002).

Where there is obvious and strong congruence between remote sensing-derived land cover classes and biological distributions, retaining sufficient examples of each land cover class has been used successfully as a conservation strategy (Ferrier, 2002). Species richness has also been correlated with habitat heterogeneity, typically represented by an index describing how many different land cover classes occur within a unit (Stoms and Estes, 1993), although elevation range within a transect has also been used (Tognelli and Kelt, 2004). While extremely heterogeneous landscapes tend to have low primary productivity, variation in NDVI can be used to represent heterogeneity (Oindo and Skidmore, 2002). Thes authors used the standard deviation of interannual (monthly) maximum NDVI values as an index of variation in vegetation structure and composition.

As the functional gradients they represent are location-specific, relationships between species distributions and spectral reflectance values are often not translatable beyond the study area in which the model data were collected. Nagendra (2001) suggests that species–spectral relationships might have to be recalculated for each new image, reducing the overall increase in efficiency of using remotely sensed variables as proxies for biodiversity. Seto *et al.* (2004) suggest that locations outside of their study area may have similar NDVI values but different effects on species richness.

One of the earliest conceptual frameworks for integrating GIS and remote sensing to map species distributions was developed by what is now the US Geological Survey's National Gap Analysis Programme (GAP; Scott *et al.*, 1993). Originally used to show 'gaps' in conservation status of land relative to the distribution of habitat for endangered or threatened vertebrate species, GAP analysis has evolved into a well-established technical and organizational framework for mapping biodiversity 'elements' (plant species, communities or habitats) (Jennings, 2000). Jennings (2000) provides an overview of the methods involved: (a) a land-cover map based on Landsat TM imagery is produced by some combination of photogrammetry, supervised and unsupervised classification, and the thematic accuracy is assessed; (b) land cover maps and GIS data (e.g. elevation, soil) are used in a SDM to produce a habitat suitability map; (c) land stewardship maps are generated, based on biodiversity management categories; and (d) biodiversity elements that are underrepresented in conservation areas are identified. These GAP vegetation and habitat suitability maps produced have subsequently been used as predictor variables for other studies (Frescino *et al.*, 2001). One incidental but no less important result of the success of GAP analysis has been the increase in GIS and remote sensing capabilities among biologists and ecologists (Jennings, 2000).

10.6 Change detection

Land-cover change detection, one of the most common uses of remotely sensed data, is an essential component of ecological monitoring (Aplin, 2005). Change detection and mapping requires land cover maps from at least two time periods (see Figure 10.1), and is possible only if changes in the surface phenomena of interest result in detectable differences in image radiance or emittance (Lunetta *et al.*, 2002). The level of mapping ranges from simple (i.e. change/no change) to complex (i.e. several ordinal change categories) as a function of the dominant land use, prospective disturbance types, management practices and study objectives (Rogan *et al.*, 2003).

The latter scenario is of particular interest to researchers involved in large-area habitat monitoring programmes, where many different types of land cover changes can occur and must be accounted for, e.g. pest infestation, logging and wildfire (Rogan and Miller, 2006). While the most common method of habitat monitoring requires the categorical comparison of independently classified maps, this approach has several drawbacks: (a) high cost (and time consumption) of mapping and re-mapping; (b) inability to detect subtle land cover modifications; and (c) categorical and positional errors in both land cover maps are compounded when compared. The production of maps depicting change can facilitate an improved understanding of both the agents of change and the biophysical linkages between surface reflectance and the change agents, e.g. NDVI can be directly linked to multitemporal changes in green vegetation cover (Lunetta *et al.*, 2002). Rogan *et al.* (2003) show the increase in accuracy of change maps when GIS variables representing topography are included in the analysis.

10.6.1 Case study: using GIS and remote sensing for large-area change detection and efficient map updating

This case study presents a hybrid change detection technique that integrates GIS and remotely sensed data for efficient change map updating. Two sophisticated, parametric and non-parametric classification techniques, generalized linear models

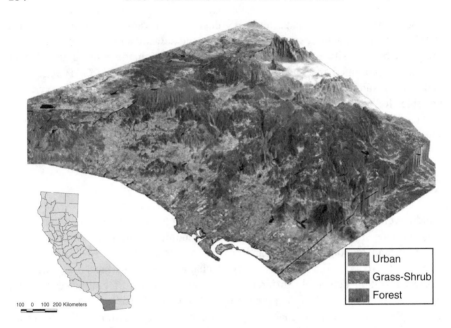

Figure 10.2 Study area location in San Diego County (California). Landsat 5 TM (2000) bands 7, 5 and 4 draped over a 30 m digital elevation model

(GLMs) and classification trees (CTs), were evaluated in terms of change map accuracy, as measured by receiver operating characteristic (ROC) plots.

10.6.1.1 Study area

San Diego County (Figure 10.2) is composed of a variety of heterogeneous land cover types, including shrub-grassland (60%), conifer and hardwood forest (12%), agriculture (6%) and urban (18%). (USFS, 2001). Mean annual precipitation is low (600 mm) and is correlated with elevation, which ranges from sea level to 1991 m. The area is currently undergoing dramatic population growth and acceler-ated and extensive land cover change due to natural and anthropogenic disturbance (Stephenson and Calcarone, 1999). These spatially and temporally diverse distur-bances result in land cover changes ranging from dramatic (e.g. wildfire burn scars, land development) to very subtle (e.g. conifer pest infestation, post-fire regenera-tion) (Rogan *et al.*, 2002). More than 50% of the county is 'Category 4' GAP land management status (i.e. unprotected habitat) (Scott *et al.*, 1993).

10.6.1.2 Data and methods

Two Landsat TM-5 images acquired 8 and 17 June 1996, and two Landsat ETM-7 images acquired 11 and 12 June 2000 (path 40/row 37 and path 39/row 37, respectively) provided coverage for San Diego County. The images were geometrically registered, resampled and normalized for atmospheric and illumination effects, using a procedure described in detail in Rogan *et al.* (2003). Six spectral variables (Kauth Thomas) and five GIS variables (slope, elevation, aspect, vegetation type and previous fire) were used to predict land cover change/no change. The Kauth Thomas variables were: change in brightness (MKT1), change in greenness (MKT2), change in wetness (MKT3), stable brightness (MKT4), stable greenness (MKT5) and stable wetness (MKT6). The modelling dataset was divided into 665 training cases used to develop the models, and 165 test cases to assess the classification accuracy of the model predictions. A full description of the field data collection protocol is provided in Rogan *et al.* (2003).

Classification trees (CTs) and generalized linear models (GLMs) were used to develop models that predicted probability or suitability of land cover change. The CT model was pruned to 11 terminal nodes, based on cross-validation (Breiman *et al.*, 1984). The GLMs were developed based on a combination of stepwise and subjective, iterative variable addition and subtraction methods, with a goal of minimizing the Akaike information criterion (AIC) statistic (Akaike, 1973; Hastie *et al.*, 2001), using all significant variables. The first GLM was based on linear relationships between change and the predictor variables (GLM_linear). A more complex GLM (GLM_poly) was tested that contained interaction terms suggested by the CT model structure, and non-linear relationships between the predictor variables and likelihood of change. More flexible non-linear relationships were explored using generalized additive models (GAM) (Franklin, 1998; Miller, 2005) and, where appropriate, were specified as polynomials (up to third order) or piecewise linear terms.

10.6.1.3 *Results*

CT models consist of a series of hierarchical binary splits, the thresholds of which are selected to maximize homogeneity in the two resulting splits. Therefore, the order in which variables are used can be an indication of their relative importance, although scale is also a factor (broad-scale variables are generally used before fine-scale variables). Figure 10.3 shows the CT models for training and test data. Rectangles represent terminal nodes with the majority class (1 = change, 0 = no change), and the fraction below is the number of cases misclassified/total cases (see Rogan *et al.*, 2003).

MKT2 (change in greenness) was the lead split, followed by vegetation, MKT3, MKT6, slope, fire and elevation. The right branch of the tree is associated with increases in the magnitude of greenness values between the two image acquisition dates. It is interesting to observe that this greenness branch was also associated with MKT6 (stable wetness). Due to acute changes in average precipitation levels between the image acquisition dates, soil–plant moisture content likely changed substantially. Recent research has demonstrated the utility of including stable

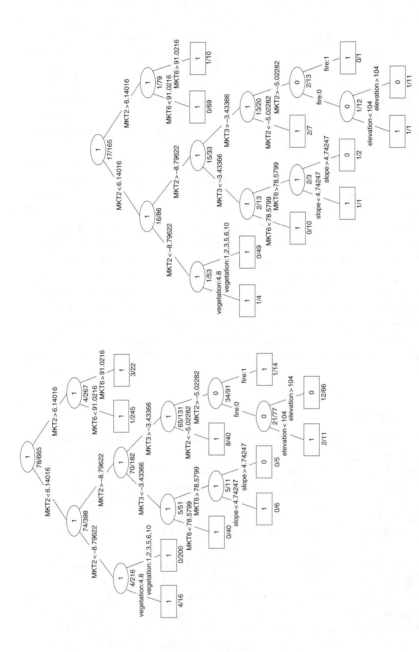

Figure 10.3 Pruned classification tree models for training (A) and test (B) data. Rectangles show terminal nodes with the majority classification (1, change; 0, no change). Fractions below the terminal nodes give misclassification rate (number incorrectly classified/total number of observations that occur at that node)

features with change features in change mapping, because stable features permit discrimination among more subtle change classes (Cohen and Fiorella, 1998; Rogan *et al.*, 2003). This is because different land cover changes can produce similar spectral signatures in measurement space (Cohen and Fiorella, 1998).

The left branch of the tree is associated with landscape changes related to suburbanization, land clearing and wildfires. These changes can be manifested spectrally, and produce substantial disturbance-specific changes in soil brightness, green vegetation cover and soil–plant moisture content. The most important spectral variables associated with disturbance were MKT2 (change in greenness), MKT (change in greenness) and MKT6 (stable wetness). These changes are also correlated with topographic variables. Gradual slopes ($< 4.7°$) were associated with change (most likely development), as were low elevations (< 104 m). This evidence compliments the association between changes in greenness, slope and elevation, and reveals a scenario of decreases in green vegetation cover and soil–plant moisture at low elevation levels and shallow slope gradients in San Diego County for the time period examined. While such a scenario is not necessarily surprising to local ecological experts and landscape planners, the results of this case study provide spatially explicit information on the actual locations of landscape changes in association with myriad disturbance agents and events. Currently, this information can only be revealed in this spatially continuous context at a 30 m minimum mapping unit, using the integration of remotely sensed data, environmental variables and statistical models in a GIS.

Table 10.3 illustrates how the ability to specify non-linear relationships improves the GLM's ability to predict probability of change. Figure 10.4 shows the generalized additive model (GAM) plots of the smoothed effect of the predictor variables on change. Elevation, MKT1, MKT2 and MKT3 all appear to have non-linear effects on probability of change. When specified as polynomials in the GLM, all four variables except MKT3 improved the model fit (Table 10.3). MKT2 is also the most important variable in GLM_poly, and was the most important non-categorical variable in GLM_linear. GLM_poly has a lower AIC statistic, indicating that it is a parsimonious model that fits the data much better than GLM_linear.

Table 10.3 Model AIC and reduction in deviance for each variable used in GLM

Model	Fire	Aspect	Elevation	Vegetation	MKT1	MKT2	MKT3	MKT4	MKT2: MKT3
GLM_linear AIC = 419	53	7	< 1*	27	2*	17	9	1*	–
GLM_poly AIC = 223	53	7	16(2nd)	22	28(3rd)	114(2nd)	10 (3rd)	16(3rd)	16

Numbers in parentheses show polynomial order.
*Variable was insignificant at $p < 0.01$.

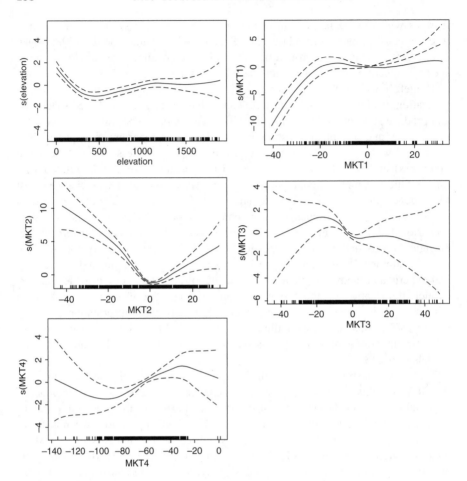

Figure 10.4 Response curves based on GAMs for predictor variables that were specified to have non-linear relationships with change

Model accuracy was assessed using test data and the area under the curve (AUC) of receiver-operating characteristic (ROC) plots. ROC plots are a threshold- and prevalence-independent metric used to measure how well a model can discriminate between two outcomes: AUC = 0.5 shows no ability, while AUC = 1.0 shows perfect ability (Fielding and Bell, 1997). The non-linear GLM also had higher model accuracy than the linear GLMs. (Figure 10.5).

The GAMs in Figure 10.4 show the non-linear relationships between probability of change and elevation, MKT1, MKT2, MKT3 and MKT4. Elevation is strongly inversely related to change in San Diego County, as the urban development occurs in the lower elevation area near the coast. MKT variables are expected to have a non-monotonic relationship with change, as they show increased values (> zero)

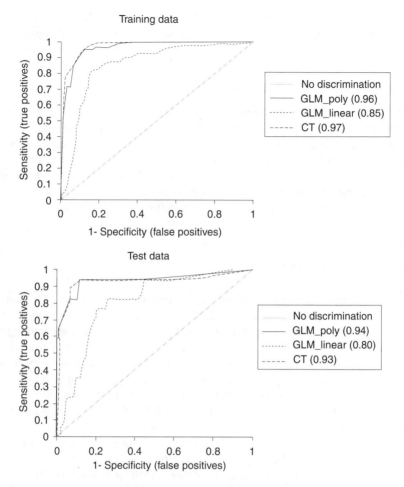

Figure 10.5 ROC plots for all three models show that the more flexible models (GLM_poly, CT) have higher accuracy than the linear GLM

and decreased values ($<$ zero) in brightness, greenness and wetness. The GAM plot of elevation shows a distinct threshold at 400 m, above which change is not likely to occur, and below which probability of change increases as elevation decreases. This particular threshold does not appear in the classification tree (although the threshold of 104 m is used in a low split), most likely because elevation is correlated with many of the MKT variables, particularly change in greenness (MKT2), and this effect is captured when these variables are used.

Classification accuracy based on the training data provides an optimistic view of model accuracy: AUC for the CT and GLM_poly models were both high (0.97 and 0.96, respectively) and not significantly different. Classification accuracy using the test data was also high for the CT and GLM_poly models (AUC = 0.93 and 0.94, respectively). Accuracy for the GLM_linear model was lower and significantly different for both training and test data (AUC = 0.85 and 0.80, respectively).

10.6.1.4 Case study discussion

A comparison of state-of-the-art mapping algorithms was performed to detect land cover change in San Diego County, California. Non-linear relationships between the spectral and GIS variable to change/no change were important, as was the combination of GIS and remotely sensed predictor variables. The success of the non-linear methods is related to the evidence (from GAMs and linear GLMs) that land cover modifications are complex and have multiple causative effects, and are therefore not amenable to traditional linear modelling approaches.

10.7 Conclusions

An understanding of the environmental factors that determine species distributions (type, abundance, level of diversity) has always been of great interest in ecological research, but its importance has increased along with interest in studying the consequences of changing environmental conditions. The ability to map, model and monitor these distributions is dependent upon the ability to collect, manage and analyse data that adequately describe them. GIS and remote sensing have become indispensable tools in this regard, providing increasingly more ecologically relevant data at higher spatial and temporal resolution, as well as the methods to derive more information from the data and to analyse them statistically.

A more extensive integration of GIS and remote sensing for ecological mapping and monitoring has yet to be fully realized. Much progress has been made in SDM based on GIS data and methods, but investigation into the potential predictive ability of remotely sensed variables, beyond land cover and NDVI, has lagged behind. We see the following as areas in which integration of GIS and remote sensing can have significant effects on ecological research:

- Use of continuous or gradual properties of spectral data should be high priority in SDM; functional gradients can enhance models based on primary and direct gradients.

- More attention should be placed on the ability to consistently characterize biological diversity at multiple scales, using remotely sensed and GIS data.

- In addition to using the reflectance values, the unique synoptic perspective of remote sensing for quantifying landscape characteristics should also be more fully studied.

- Change detection models should explore more flexible statistical methods and include ancillary GIS data.

Acknowledgements

The authors would like to thank Victor Mesev, Ben Gilmer (West Virginia University) and Timothy Currie (Clark University) for reviews and assistance with this manuscript. Funding from the National Science Foundation (award 0451486 is gratefully acknowledged.

References

Akaike, H. (1973) Information theory and an extension of the maximum likelihood. In 2nd International Symposium of Information Theory, Budapest, Hungary.

Aplin, P. (2005) Remote sensing: ecology. *Progress in Physical Geography* **29**, 104–113.

Ardo, J. (1992) Volume quantification of coniferous forest compartments using spectral radiance recorded by Landsat thematic mapper. *International Journal of Remote Sensing* **13**, 1779–1786.

Austin, G., Thomas, C., Houston, D. and Thompson, D. B. A. (1996) Predicting the spatial distribution of buzzard *Buteo buteo* nesting areas using a geographical information system and remote sensing. *Journal of Applied Ecology* **33**, 1541–1550.

Austin, M. and Gaywood, M. (1994) Current problems of environmental gradients and species response curves in relation to continuum theory. *Journal of Vegetation Science* **5**, 473–482.

Austin, M. P. (1987) Models for the analysis of species' response to environmental gradients. *Vegetatio* **69**, 35–45.

Austin, M. P. (2002) Spatial prediction of species distribution: an interface between ecological theory and statistical modelling. *Ecological Modelling* **157**, 101–118.

Austin, M. P. and Heyligers, P. C. (1989) Vegetation survey design for conservation: gradsect sampling of forests in north-east New South Wales. *Biological Conservation* **50**, 13–32.

Belward, A. S., Estes, J. E. and Kline, K. D. (1999) The IGBP–DIS global 1 km landcover data set DIScover: a project overview. *Photogrammetric Engineering and Remote Sensing* **65**, 1013–1020.

Bio, A., Alkemade, R. and Barendregt, A. (1998) Determining alternative models for vegetation response analysis: a non-parametric approach. *Journal of Vegetation Science* **9**, 5–16.

Breiman, L., Freedman, J., Olshen, R. and Stone, C. (1984) *Classification and Regression Trees*. Wadsworth International Group: Belmont, CA, USA.

Busby, J. R. (2002) Biodiversity mapping and modelling. In Skidmore, A. (ed.), *Environmental Modelling with GIS and Remote Sensing*. Taylor and Francis: London, UK, 145–165.

Cairns, D. M. (2001) A comparison of methods for predicting vegetation type. *Plant Ecology* **156**, 3–18.

Chen, J. M. and Black, T. A. (1992) Defining leaf area index for non-flat leaves. *Plant, Cell and Environment* **15**, 421–429.

Chen, J. M. and Guilbeault, M. (1996) Evaluation of vegetation indices for deriving biophysical parameters of boreal forests. In Proceedings of the 26th International Symposium on
Remote Sensing of Environment, Vancouver, Canada.

Cihlar, J., Latifovic, R., Beaubien, J., Trishchenko, A., Chen, J. and Fedosejevs, G. (2003)
National scale forest information extraction from coarse information extraction from coarse
resolution satellite data, Part 1. Data processing and mapping land cover types. In Wulder,
M. and Franklin, S. (eds), *Remote Sensing of Forest Environments: Concepts and Case
Studies*. Kluwer Academic: Dordrecht, The Netherlands, 337–357.

Cohen, W. B. and Fiorella, M. (1998) Comparison of methods for detecting conifer change
with Thematic Mapper imagery. In Lunetta, R. S. and Elvidge, C. D. (eds), *Remote
Sensing Change Detection: Environmental Monitoring Methods and Applications*. Ann
Arbor Press: Chelsea, MI, USA, 89–102.

Cohen, W. B. and Justice, C. O. (1999) Validating MODIS terrestrial ecology products:
linking *in situ* and satellite measurements. *Remote Sensing of Environment* **70**, 1–3.

Cohen, W. B., Spies, T. and Fiorella, M. (1995) Estimating the age and structure of forests
in a multi-ownership landscape of western Oregon, USA. *International Journal of Remote
Sensing* **16**, 721–746.

Cohen, W. B. and Spies, T. A. (1992) Estimating structural attributes of Douglas fir/western
hemlock forest stands from LANDSAT and SPOT imagery. *Remote Sensing of Environment* **41**, 1–17.

Coops, N. C. and Catling, P. C. (2002) Prediction of the spatial distribution and relative
abundance of ground-dwelling mammals using remote sensing imagery and simulation
models. *Landscape Ecology* **17**, 173–188.

Coppin, P., Jonckheere, I., Nackaerts, K., Muys, B. and Lambin, E. (2004) Digital change
detection methods in ecosystem monitoring: a review. *International Journal of Remote
Sensing* **25**, 1565–1596.

Cosentino, M. J., Woodcock, C. E. and Franklin, J. (1981) Scene analysis for wildland fire-
fuel characteristics in a Mediterranean climate. In Proceedings of the Fifteenth International
Symposium on Remote Sensing of Environment, Ann Arbor, MI, USA.

Crist, E. P. and Ciccone, R. C. (1984) Application of the tasseled cap concept to simulated
Thematic Mapper data. *Photogrammetric Engineering and Remote Sensing* **50**, 343–352.

Cumming, G. S. (2002) Comparing climate and vegetation as limiting factors for species
ranges of African ticks. *Ecology* **83**, 255–268.

Curran, P. J. (2001) Remote sensing: using the spatial domain. *Environmental and Ecological
Statistics* **8**, 331–344.

Curran, P. J., Milton, E. J., Atkinson, P. J. and Foody, G. M. (1998) Remote sensing: from
data to understanding. In Longley, P. A., Brooks, S. M., McDonnell, R. and Macmillan,
B. (eds), *Geocomputation: A Primer*. Wiley: Chichester, UK, 33–59.

Danson, F. M. and Curran, P. J. (1993) Factors affecting the remotely sensed response of
coniferous forest plantations. *Remote Sensing of the Environment* **63**, 59–67.

Davis, F. W. and Roberts, D. A. (2000) Stand structure in terrestrial ecosystems. In Sala,
O. E., Jackson, R. B., Mooney, H. A. and Howarth, R. W. (eds), *Methods in Ecosystem
Science*. Springer: New York, NY, USA, 7–30.

de Leeuw, J., Ottichilo, W. K., Toxopeus, A. G. and Prins, H. H. T. (2002) Application of
remote sensing and geographic information systems in wildlife mapping and modelling.
In Skidmore, A. (ed.), *Environmental Modelling with GIS and Remote Sensing*. Taylor
and Francis: London, UK, 121–144.

Debinski, D. M., Kindscher, K. and Jakubauskas, M. E. (1999) A remote sensing and GIS-based model of habitats and biodiversity in the Greater Yellowstone Ecosystem. *International Journal of Remote Sensing* **20**, 3281–3291.

Diamond, J. (1988) Factors controlling species diversity: overview and synthesis. *Annals of the Missouri Botanical Garden* **75**, 111–129.

Dirnböck, T., Dullinger, S., Gottfried, M., Ginzler, C. and Grabherr, G. (2003) Mapping alpine vegetation based on image analysis, topographic variables and canonical correspondence analysis. *Applied Vegetation Science* **6**, 85–96.

Edwards, T. C., Cutler, D. R., Zimmermann, N., Geiser, L. and Alegria, J. (2005) Model-based stratifications for enhancing the detection of rare ecological events. *Ecology* **86**, 1081–1090.

EPA (Environmental Protection Agency). (1998) Ecological Research Strategy. EPA/600/R-98/086, June 1998. EPA: Washington, DC, USA; 130 Available at: http://www.epa.gov/ORD/WebPubs/final

Ferrier, S. (2002) Mapping spatial pattern in biodiversity for regional conservation planning: where to from here? *Systematic Biology* **51**, 331–363.

Ferrier, S., Drielsma, M., Manion, G. and Watson, G. (2002b) Extended statistical approaches to modelling spatial pattern in biodiversity in north-east New South Wales. II. Community-level modelling. *Biodiversity and Conservation* **11**, 2309–2338.

Fielding, A. H. and Bell, J. F. (1997) A review of methods for the assessment of prediction errors in conservation presence/absence models. *Environmental Conservation* **24**, 38–49.

Florinsky, I. V. (1998) Combined analysis of digital terrain models and remotely sensed data in landscape investigations. *Progress in Physical Geography* **22**, 33–60.

Franklin, J. (1995) Predictive vegetation mapping: geographic modelling of biospatial patterns in relation to environmental gradients. *Progress in Physical Geography* **19**, 474–499.

Franklin, J. (1998) Predicting the distributions of shrub species in California chaparral and coastal sage communities from climate and terrain-derived variables. *Journal of Vegetation Science* **9**, 733–748.

Franklin, J. (2002) Enhancing a regional vegetation map with predictive models of dominant plant species in chaparral. *Applied Vegetation Science* **5**, 135–146.

Franklin, J., Logan, T. L., Woodcock, C. E. and Strahler, A. (1986) Coniferous forest classification and inventory using Landsat and digital terrain data. *IEEE Transactions on Geoscience and Remote Sensing* **24**, 139.

Franklin, J., McCullough, P. and Gray, C. (2000) Terrain variables used for predictive mapping of vegetation communities in Southern California. In Wilson, J. and Gallant, J. (eds), *Terrain Analysis: Principles and Applications*. Wiley: New York, NY, USA, 331–353.

Franklin, J. and Turner, D. L. (1992) The application of a geometric optical canopy reflectance model to semi-arid shrub vegetation. *IEEE Transactions on Geoscience and Remote Sensing* **30**, 293–301.

Franklin, S. E. (2001) *Remote Sensing for Sustainable Forest Management*. Lewis: Boca Raton, FL, USA.

Frescino, T., Edwards, T. and Moisen, G. (2001) Modelling spatially explicit forest structural attributes using generalized additive models. *Journal of Vegetation Science* **12**, 15–26.

Friedl, M. A., Michaelsen, J., Davis, F. W., Walker, H. and Schimel, D. S. (1994) Estimating grassland biomass and leaf area index using ground and satellite data. *International Journal of Remote Sensing* **15**, 1401–1420.

Gholz, H. L. (1982) Environmental limits on aboveground net primary production, leaf area, and biomass in vegetation zones of the Pacific northwest. *Ecology* **63**, 469–481.

Gibson, L. A., Wilson, B. A., Cahill, D. M. and Hill, J. (2004) Spatial prediction of rufous bristlebird habitat in a coastal heathland: a GIS-based approach. *Journal of Applied Ecology* **41**, 213–223.

Gong, P. and Xu, B. (2003) Multi-spectral and multi-temporal image processing approaches: part 2. Change detection. In Wulder, M. A. and Franklin, S. E. (eds), *Methods and Applications for Remote Sensing of Forests: Concepts and Case Studies*. Kluwer Academic: Dordrecht, The Netherlands.

Goodchild, M. F. (1994) Integrating GIS and remote sensing for vegetation analysis and modelling: methodological issue. *Journal of Vegetation Science* **5**, 615–626.

Gottschalk, T. K., Huettmann, F. and Ehlers, M. (2005) Thirty years of analyzing and modelling avian habitat relationships using satellite imagery data: a review. *International Journal of Remote Sensing* **26**, 2631–2656.

Guisan, A. and Thuiller, W. (2005) Predicting species distribution: offering more than simple habitat models. *Ecology Letters* **8**, 993–1009.

Guisan, A. and Zimmermann, N. (2000) Predictive habitat distribution models in ecology. *Ecological Modelling* **135**, 147–186.

Hansen, M. C., DeFries, R. S., Townshend, J. R. G. and Sohlberg, R. (2000) Global land cover classification at 1 km spatial resolution using a classification tree approach. *International Journal of Remote Sensing* **21**, 1331–1364.

Hastie, T., Tibshirani, R. and Friedman, J. (2001) *The Elements of Statistical Learning: Data Mining, Inference and Prediction*. Springer-Verlag: New York, NY, USA.

Holdridge, L. (1947) Determination of world plant formations from simple climatic data. *Science* **105**, 367–368.

Holdridge, L. (1967) *Life Zone Ecology*. Tropical Science Centre: San Jose, Costa Rica.

Holmes, K. W., Chadwick, O. A. and Kyriakidis, P. C. (2000) Error in a USGS 30-meter digital elevation model and its impact on terrain modelling. *Journal of Hydrology* **233**, 154–173.

Homer, C., Ramsey, R. T. E. Jr and Falconer, A. (1997) Landscape cover-type modelling using a multi-scene thematic mapper mosaic. *Photogrammetric Engineering and Remote Sensing* **63**, 59–67.

Huang, C., Homer, C. and Yang, L. (2003) Regional forest land-cover characterizations using medium spatial resolution satellite data. In Wulder, M. and Franklin, S. E. (eds), *Methods and Applications for Remote Sensing of Forests: Concepts and Case Studies*. Kluwer Academic: Dordrecht, The Netherlands, 389–410.

Jakubauskas, M. E. and Price, K. P. (1997) Empirical relationships between biotic and spectral factors of Yellowstone lodgepole pine forests. *Photogrammetric Engineering and Remote Sensing* **63**: 1375–1381.

Jeganathan, P., Green, R. E., Norris, K., Vogiatzakis, I. N., Bartsch, A., Wotton, S. R., Bowden, C. G. R., Griffiths, G. H., Pain, D. and Rahmani, A. R. (2004) Modelling habitat selection and distribution of the critically endangered Jerdon's courser *Rhinoptilus bitorquatus* in scrub jungle: an application of a new tracking method. *Journal of Applied Ecology* **41**, 224–237.

Jennings, M. D. (2000) Gap analysis: concepts, methods, and recent results. *Landscape Ecology* **15**, 5–20.

Jensen, J. R. (2000) *Remote Sensing of the Environment: an Earth Resource Perspective*. Prentice-Hall: Saddle River, NJ, USA.

Jongman, R., Ter Braak, C. and Van Tongeren, O. (eds). (1995) *Data Analysis in Community and Landscape Ecology*. Cambridge University Press: Cambridge, MA, USA.

Kasischke, E. S., Melack, J. M. and Dobson, M. C. (1997) The use of imaging radars for ecological applications. *Remote Sensing of Environment* **59**, 141–156.

Kauth, R. J. and Thomas, G. S. (1976) The tasseled cap – a graphic description of the spectral-temporal development of agricultural crops as seen by Landsat. In Proceedings of the Symposium on Machine Processing of Remotely Sensed Data. Purdue University, West Lafayette, IN, USA.

Kerr, J. T. and Ostrovsky, M. (2003) From space to species: ecological applications for remote sensing. *Trends in Ecology and Evolution* **18**, 299–305.

Kessell, S. (1979) *Gradient Modelling: Resource and Fire Management*. Springer-Verlag: New York, NY, USA.

Leathwick, J. (1998) Are New Zealand's *Nothofagus* species in equilibrium with their environment? *Journal of Vegetation Science* **9**, 719–732.

Lees, B. G. and Ritman, K. (1991) Decision-tree and rule-induction approach to integration of remotely sensed and GIS data in mapping vegetation in disturbed or hilly environments. *Environmental Management* **15**, 823–831.

Liu, J. and Taylor, W. W. (eds). (2002) *Integrating Landscape Ecology into Natural Resource Management*. Cambridge University Press: Cambridge, MA, USA.

Lunetta, R. S., Alvarez, R., Edmonds, C. M., Lyon, J. G., Elvidge, C. D., Bonifaz, R. and Garcia, C. (2002) NALC/Mexico land-cover mapping results: implications for assessing landscape condition. *International Journal of Remote Sensing* **23**, 3129–3148.

Luoto, M., Kuussaari, M. and Toivonen, T. (2002a) Modelling butterfly distribution based on remote sensing data. *Journal of Biogeography* **29**: 1027–1037.

Luoto, M., Toivonen, T. and Heikkinen, R. (2002b) Prediction of total and rare plant species richness in agricultural landscapes from satellite images and topographic data. *Landscape Ecology* **17**, 195–217.

McDermid, G. J., Franklin, S. E. and LeDrew, E. F. (2005) Remote sensing for large-area habitat mapping. *Progress in Physical Geography* **29**, 449–474.

McGarigal, K. and McComb, W. C. (1995) Relationships between landscape structure and breeding birds in the Oregon Coast Range. *Ecological Monographs* **65**, 235–260.

Meentemeyer, R., Moody, A. and Franklin, J. (2001) Landscape-scale patterns of shrub-species abundance in California chaparral: the role of topographically mediated resource gradients. *Plant Ecology* **156**, 19–41.

Michaelsen, J., Schimel, D., Friedl, M., Davis, F. W. and Dubayah, R. C. (1994) Regression tree analysis of satellite and terrain data to guide vegetation sampling and surveys. *Journal of Vegetation Science* **5**, 673–686.

Miller, J. (2005) Incorporating spatial dependence in predictive vegetation models: residual interpolation methods. *The Professional Geographer* **57**, 169–184.

Moore, I., Grayson, R. and Ladson, A. (1991) Digital terrain modelling: a review of hydrological, geomorphological, and biological applications. *Hydrological Processes* **5**, 3–30.

Müller, F. (1998) Gradients in ecological systems. *Ecological Modelling* **108**, 3–21.

Nagendra, H. (2001) Using remote sensing to assess biodiversity. *International Journal of Remote Sensing* **22**, 2377–2400.

Oindo, B. and Skidmore, A. (2002) Interannual variability of NDVI and species richness in Kenya. *International Journal of Remote Sensing* **23**, 285–298.

Osborne, P. E., Alonso, J. C. and Bryant, R. G. (2001) Modelling landscape-scale habitat use using GIS and remote sensing: a case study with great bustards. *Journal of Applied Ecology* **38**, 458–471.

Oza, M. P., Strivastava, V. K. and Devaiah, P. K. (1996) Estimating tree volume in tropical dry deciduous forest from Landsat TM data. *Geocarto International* **114**, 33–39.

Parra, J. L., Graham, C. C. and Freile, J. F. (2004) Evaluating alternative data sets for ecological niche models of birds in the Andes. *Ecography* **27**, 350–360.

Pearson, R., Dawson, T. and Liu, C. (2004) Modelling species distributions in Britain: a hierarchical integration of climate and land-cover data. *Ecography* **27**, 285–298.

Pettorelli, N., Olak Vik, J., Mysterud, A., Gaillard, J. -M., Tucker, C. J. and Stenseth, N. C. (2005) Using the satellite-derived NDVI to assess ecological responses to environmental change. *Trends in Ecology and Evolution* **20**, 503–510.

Ramsey, R. D., Falconer, A. and Jensen, J. R. (1995) The relationship between NOAA-AVHRR NDVI and ecoregions in Utah. *Remote Sensing of Environment* **53**, 188–198.

Roberts, D. A., Batista, G. T., Pereira, J., Waller, E. K. and Nelson, B. W. (1998) Change identification using multitemporal spectral mixture analysis applications in eastern Amazonia. In Lunetta, R. S. and Elvidge, C. D. (eds), *Remote Sensing Change Detection Environmental Monitoring Methods and Applications.* Ann Arbor Press: Chelsea, MI, USA, 318.

Rogan, J. and Chen, D. M. (2004) Remote sensing technology for mapping and monitoring land cover and land use change. *Progress in Planning* **61**, 301–325.

Rogan, J., Franklin, J. and Roberts, D. A. (2002) A comparison of methods for monitoring multitemporal vegetation change using Thematic Mapper imagery. *Remote Sensing of Environment* **80**, 143–156.

Rogan, J. and Miller, J. (2006) Integrating GIS and remote sensing for mapping forest disturbance and change. In Wulder, M. A. and Franklin, S. E. (eds), *Understanding Forest Disturbance and Spatial Pattern: Remote Sensing and GIS Approaches.* CRC Press (Taylor & Francis): Boca Raton, FL, USA

Rogan, J., Miller, J., Stow, D., Franklin, J., Levien, L. and Fischer, C. (2003) Land cover change mapping in California using classification trees with Landsat TM and ancillary data. *Photogrammetric Engineering and Remote Sensing* **69**, 793–804.

Rollins, M. G., Keane, R. and Parsons, R. (2004) Mapping fuels and fire regimes using remote sensing, ecosystem simulation, and gradient modelling. *Ecological Applications* **14**, 75–95.

Rushton, S. P., Ormerod, S. J. and Kerby, G. (2004) New paradigms for modelling species distributions? *Journal of Applied Ecology* **41**, 193–200.

Scott, J., Davis, F., Csuti, B., Noss, R., Butterfield, B., Groves, C., Anderson, H., Caicco, S., D'Erchia, F., Edwards, T., Ulliman, J. and Wright, R. (1993) Gap analysis: a geographic approach to protection of biological diversity. *Wildlife Monographs* **123**, 1–41.

Scott, J. M., Heglund, P. J., Morrison, M. L., Haufler, J. B., Raphael, M. G., Wall, W. A. and Samson, F. B. (eds). (2002) *Predicting Species Occurrences: Issues of Accuracy and Scale.* Island Press: Washington, DC, USA.

Seto, K., Fleishman, E., Fay, J. P. and Betrus, C. (2004) Linking spatial patterns of bird and butterfly species richness with Landsat TM-derived NDVI. *International Journal of Remote Sensing* **25**, 4309–4324.

Skidmore, A., Oindo, B. and Said, M. (2003) Biodiversity assessment by remote sensing. In 30th International Symposium on Remote Sensing of the Environment: Information for Risk Management and Sustainable Development, Honolulu, HI, USA.

Steininger, M. K. (2000) Satellite estimation of tropical secondary forest above-ground biomass: data from Brazil and Bolivia. *International Journal of Remote Sensing* **21**, 1139–1157.

Stephenson, J. R. and Calcarone, G. (1999) Southern California mountains and foothills assessment: habitat and species conservation issues. General Technical Report GTR-PSW-172. US Department of Agriculture, Pacific Southwest Research Station, Forest Service: Albany, CA, USA; 402 pp.

Stoms, D. M. and Estes, J. E. (1993) A remote sensing research agenda for mapping and monitoring biodiversity. *International Journal of Remote Sensing* **14**, 1839–1860.

Stoms, D. M. and Hargrove, W. (2000) Potential NDVI as a baseline for monitoring ecosystem functioning. *International Journal of Remote Sensing* **21**, 401–407.

Stone, T., Schlesinger, P., Houghton, T. and Woodwell, G. (1994) A map of the vegetation of South America based on satellite imagery. *Photogrammetric Engineering and Remote Sensing* **60**, 541–551.

Stow, D. (1995) Monitoring ecosystem response to global change: multitemporal remote sensing analysis. In Moreno, J. and Oechel, W. (eds), *Anticipated Effects of a Changing Global Environment in Mediterranean Type Ecosystems*. Springer-Verlag: New York, NY, USA.

Suárez-Seoane, S., Osborne, P. E. and Alonso, J. C. (2002) Large-scale habitat selection by agricultural steppe birds in Spain: identifying species-habitat responses using generalized additive models. *Journal of Applied Ecology* **39**: 755–771.

Suárez-Seoane, S., Osborne, P. E. and Rosema, A. (2004) Can climate data from METEOSAT improve wildlife distribution models? *Ecography* **27**, 629–636.

Teixeira, J. and Arntzen, J. W. (2002) Potential impact of climate warming on the distribution of the Golden-striped salamander, *Chioglossa lusitanica*, on the Iberian Peninsula. *Biodiversity and Conservation* **11**, 2167–2176.

Thuiller, W., Araújo, M. B. and Lavorel, S. (2004) Do we need land-cover data to model species distributions in Europe? *Journal of Biogeography* **31**, 353–361.

Thuiller, W., Vayreda, J., Pino, J., Sabate, S., Lavorel, S. and Gracia, C. (2003) Large-scale environmental correlates of forest tree distributions in Catalonia (NE Spain). *Global Ecology and Biogeography* **12**, 313–325.

Tognelli, M. F. and Kelt, D. A. (2004) Analysis of determinants of mammalian species richness in South America using spatial autoregressive models. *Ecography* **27**, 427–436.

Travis, J. M. J. (2003) Climate change and habitat destruction: a deadly anthropogenic cocktail. *Proceedings of the Royal Society of London B (Biology)* **270**, 467–473.

Treitz, P. and Rogan, J. (2004) Remote sensing for mapping and monitoring land cover and land use change: an introduction. *Progress in Planning* **61**, 269–279.

Tucker, C. J. (1979) Red and photographic infrared linear combinations for monitoring vegetation. *Remote Sensing of the Environment* **8**, 127–150.

Turner, W., Spector, S., Gardiner, N., Fladeland, M., Sterling, E. and Steininger, M. (2003) Remote sensing for biodiversity science and conservation. *Trends in Ecology and Evolution* **18**, 306–314.

Urban, D. L. (2002) Tactical monitoring of landscapes. In Liu, J. and Taylor, W. W. (eds), *Integrating Landscape Ecology into Natural Resource Management*. Cambridge University Press: Cambridge, MA, USA, 294–311.

USFS (2001) US Forest Service, 2001. CalVeg Geobook. US Department of Agriculture, Forest Service, Region 5 Remote Sensing Lab: Sacramento, CA, USA [CD-ROM].

Van Niel, K., Laffan, S. and Lees, B. (2004) Effect of error in the DEM on environmental variables for predictive vegetation modelling. *Journal of Vegetation Science* **15**, 747–756.

Venier, L. A., Pearce, J., McKee, J. E., McKenney, D. W. and Niemi, G. J. (2004) Climate and satellite-derived land cover for predicting breeding bird distribution in the Great Lakes Basin. *Journal of Biogeography* **31**, 315–331.

Waser, L. T., Stofer, S., Schwarz, M., Kuchler, M., Ivits, E. and Scheidegger, C. (2004) Prediction of biodiversity – regression of lichen species richness on remote sensing data. *Community Ecology* **5**, 121–133.

White, A. B., Kumar, P. and Tcheng, D. (2005) A data-mining approach for understanding topographic control in climate-induced inter-annual vegetation variability over the United States. *Remote Sensing of Environment* **98**: 1–20.

Whittaker, R. (1973) Direct gradient analysis. In Whittaker, R. W. (ed.), *Ordination and Classification of Communities*. Junk: The Hague, The Netherlands, 9–45.

Wiens, J. A., Van Horne, B. and Noon, B. R. (2002) Integrating landscape structure and scale into natural resource management. In Liu, J. and Taylor, W. W. (eds), *Integrating Landscape Ecology into Natural Resource Management*. Cambridge University Press: Cambridge, MA, USA, 23–67.

Woodcock, C. E., Macomber, S. and Kumar, L. (2002) Vegetation mapping and monitoring. In Skidmore, A. (ed.), *Environmental Modelling with GIS and Remote Sensing*. Taylor and Francis: London, UK, 97–120.

Woodcock, C. E., Macomber, S. A., Pax-Lenney, M. and Cohen, W. B. (2001) Monitoring large areas for forest change using Landsat: generalization across space, time and Landsat sensors. *Remote Sensing of Environment* **78**, 194–203.

Wulder, M. (1998) Optical remote sensing techniques for the assessment of forest inventory and biophysical parameters. *Progress in Physical Geography* **22**, 449–476.

Wulder, M. A., Hall, R. J., Coops, N. C. and Franklin, S. E. (2004) High spatial resolution remotely sensed data for ecosystem characterization. *BioScience* **54**, 511–521.

Zimmermann, N., Edwards, T., Moisen, G., Frescino, T. and Blackard, J. (2007) Remote sensing-based predictors improve distribution models of rare, early successional and broadleaf tree species in Utah. *Journal of Applied Ecology* (OnlineEarly Articles).

11

Remote sensing and GIS for ephemeral wetland monitoring and sustainability in southern Mauritania

Tara Shine* and Victor Mesev†

*Environment and Development Consultant, 127 The Meadows, Belgooly, Co. Cork, Ireland

†Department of Geography, Florida State University, Tallahassee, FL, USA

11.1   Introduction

Applications of remotely sensed data for monitoring inland, arid-zone ephemeral wetlands are less conspicuous than studies of their more permanent temperate counterparts. This imbalance in the literature is unfortunate given that, as in the Sahel, ephemeral wetlands are critical life-bloods for sensitive ecological habits and the near-subsistence survival of the local population. Their monitoring by remote sensor data would result in a greater level of predictability and instil a stronger sense of confidence in their sustainable management. In other words, a temporal application of remote sensor data would provide a consistent yardstick with which to measure the annual and seasonal variability of the areal extent of transient wetlands and thus monitor changes to natural ecosystems and human lifestyles that directly depend on them.

11.1.1   Ephemeral wetlands

By definition, the size and duration of arid-zone ephemeral wetlands are highly variable. Ephemeral wetlands exist at a variety of temporal scales and are almost completely reliant on seasonal precipitation and run-off (Whittaker, 1998). Across the Sahel they represent patches of aquatic habitats for resident birds, mammals, amphibians and reptiles, including newly-documented relict populations of the Nile crocodile (*Crocodylus niloticus*). These wetlands act as crucial stop-over sites and wintering grounds for migratory water birds (cf. Roux and Jarry, 1984; Mullié *et al.*, 1994; Kingsford, 1995; Simmons *et al.*, 1999; Shine *et al.*, 2001). In addition, the ephemeral wetlands of the Sahel sustain local livelihoods by providing water and grazing for livestock, humid soil for flood recession agriculture, wild foods and medicinal plants, forest-derived products and habitats for wild animals and fish. These human-based activities are centred on traditional multi-use systems, which have existed in equilibrium with the wetlands for centuries but more recently have been stretched to their limits by increases in population numbers and the switch to more sedentary lifestyles. Furthermore, federal policies on wetland resources have been overly exploitative; for instance, agriculture associated with the ephemeral wetlands has been greatly intensified through draining and levelling and by the introduction of mechanized tools and the use of modern fertilizers and pesticides. Yet despite considerable federal investment there is growing evidence to suggest that these modern systems are less productive than the traditional multi-use systems they were designed to replace (Shine, 2002). Many alternatives are being currently explored by central government, aid agencies and NGOs, but all policies aimed at the sustainable preservation of fragile ecosystems must first address the urgent need to improve base data and provide a means with which to maintain consistent long-term inventories.

11.1.2 Remote sensing of ephemeral wetlands

A press release, dated 16 February 2005, from the European Commission's Directorate General, Joint Research Centre (DG JRC), underlined continued support for the implementation of an environmental information system based on satellite sensor data for the monitoring of Africa's natural resources, and specifically 'the location and timing of water resource replenishment and exhaustion' using 'land-resource maps' (EC, 2005). The initiative is in tune with the prospects of a programme on global monitoring for environment and security (GMES) by the European Space Agency due by 2008 to '. . . facilitate and foster the operational provision of quality data, information and knowledge', and where the DG JRC would be charged with extending the application to Africa. Such calls for a coordinated collection of data on natural resources in the developing world are long overdue. In the case of the Sahel, the lack of detailed and up-to-date information on the geographic distribution of ephemeral wetlands is not surprising. Sparse rural populations, poor infrastructure and an underdeveloped economic base in many countries with low levels of GDP are not particularly conducive to persuading governments to improve upon existing collections of unstructured inventories, generally composed of sporadically revised maps and site-specific visual observations. This is in stark contrast to national policies in the developed world, where detailed, consistent and multitemporal remotely sensed data form the cornerstones of many environmental monitoring projects (*inter alia*, Stewart *et al.*, 1986; Williams and Lyon, 1997; Harvey and Hill, 2001; Dwiveldi and Sreenivas, 2002). Furthermore, the costs of remote sensor data for governments in the developing world represent proportionally heavier investment commitments and, along with equally expensive computer infrastructure, expertise and institutional inertia, are major impediments to routine pragmatic implementation. However, and as recognized by the DG JRC, by not adopting remote sensing technology many developing countries are forfeiting a valuable vehicle with which to facilitate reliable, consistent and frequent snapshots of their ever-fluctuating natural resource base, particularly their ecologically sensitive wetland habitats.

Given the geographic vastness and relative inaccessibility of many Sahelian countries, remote sensing is by far the most practical means by which to collect such systematic multitemporal information on wetland variability, and as such is a valuable precursor to effective management policies that could harmonize the balance between ecological preservation and economic opportunity. Although works by Jensen *et al.* (1986, 1987), Rutchey and Vilcheck (1994), Chopra *et al.* (2001), Harvey and Hill (2001), Manson *et al.* (2001) and Shaikh *et al.* (2001) have all specifically addressed inland, arid-zone wetlands, none has examined seasonal, annual and long-term changes of small, more ecologically sensitive, ephemeral wetlands, such as those in the Sahel. In addition to financial restrictions, the long-term monitoring of scarce resources, such as ephemeral wetlands, is also heavily dependent on the availability of consistent remotely sensed data. In practice, many projects are compiled from a compromise of space-borne, air-borne and field-based data, but mixtures need not necessarily devalue results as long as the minimum scale and maximum range of the data are within the practical requirements of the application (cf. Lee and Lunetta, 1995; Manson *et al.*, 2001; Harvey and Hill, 2001; Lyon, 2001). This rule is particularly important in economically underdeveloped and isolated areas such as the Sahel, where a full range of data types is not normally available.

However, if economic and availability barriers can be breached, there are very few technical reasons why air-borne and space-borne data should not form the basis for routine wetland monitoring in arid parts of the world. Current spectral- and spatial-based methodologies are well capable of producing high identification and taxonomic accuracies. Techniques such as the Normalized difference vegetation index, the Normalized difference water index, tasselled cap, and standard per-pixel classifications are frequently implemented with remote sensor data for measuring biomass concentrations of wetlands in temperate landscapes (Jensen *et al.*, 1986; Ceccato *et al.*, 2002; Maselli and Rembold, 2002). Applied to arid-zone wetlands, the same techniques could measure the abrupt spectral and spatial contrast between, on the one hand, water and biomass land covers associated with ephemeral wetlands and, on the other hand, almost permanent arid surroundings (Howland, 1980; Lyon, 2001). Admittedly, ephemeral wetlands are much shallower

than their more permanent counterparts, which tends to produce spectral representations that are combinations of variable quantities of biomass, turbidity and basin
floors. This in turn leads to lower classification accuracies than those representing
more permanent and deeper water. Assuming satisfactory spectral differentiation,
semantic descriptions of ephemeral wetlands include, amongst others, the MedWet
(Farinha *et al.* 1996), the Martin (Stewart *et al.* 1980) or the UFSWS (Cowardin
and Myers, 1974) systems; the latter includes descriptions such as 'unconsolidated bottom, unconsolidated shore, emergent wetland and forested wetland'. The
Anderson *et al.* (1976) system is more harmonious with information extracted from
remote sensor data, although the temporal criterion as part of the definition of
wetlands, '. . . areas where the water table is at, near or above the land surface for a
significant part of most years . . . ' does not strictly fit the ephemeral nature of arid
zone wetlands. Indeed, all of the established classification schemes are deficient
in some ways for labelling wetlands that are fleeting and highly sensitive to small
variations in rainfall patterns.

Bearing in mind the urgent need for long-term monitoring and inventory-building
of ephemeral wetlands and the technical feasibility of manipulating remote sensor
data to fulfil these objectives, this chapter outlines a multisource methodology using
a range of remote sensor data and GPS readings over a 44 year period. Such a lengthy
timescale invariably diminishes consistency in data across source and scale, relying
instead on availability; in our case, aerial photographs captured in the 1950s, Landsat
TM images taken in 1984 and 1985, and GPS readings collected in 1999 and 2000.
Naturally, such variety can introduce spatial discrepancies, especially as in the case of
the Sahel, where there are few identifiable features with which to geo-rectify multi-
scale and multitemporal images (Manson *et al.*, 2001). Nevertheless, the available data
are of high quality, of the same geographical area and, more importantly, the coarsest
spatial resolution; that of the Landsat TM image at 30 m (< 0.1 ha) is well within the
scale requirements for measuring the extent of wetland areas (Jensen *et al.*, 1986).
Furthermore, the rather *ad hoc* combination of these data replicates, to a large extent,
current methodologies implemented by financially limited governments in the developing world. The precise methodology involves the areal delineation of ephemeral
wetland boundaries from these remote sensing sources and GPS readings using a
combination of visual image interpretation, normalized difference vegetation index
(NDVI), tasselled cap and water body spectral segmentation. The results are used
to assess seasonal and annual wetland variability with respect to sustainable habitat
management and economic development of inland ephemeral wetlands in an arid
environment.

11.2   Ephemeral wetlands in Mauritania

For an example of Sahelian ephemeral wetlands, we focus on the country of
Mauritania in western Africa, a sparse country of just over two million people

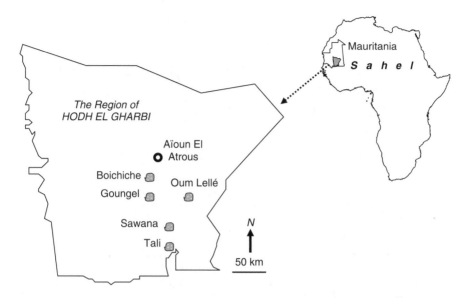

Figure 11.1   Location of the five ephemeral wetlands in the Hodh el Gharbi region, southern Mauritania, Africa

scattered across approximately one million km^2 of predominantly arid land. Our methodology is centred on five freshwater ephemeral wetlands (Tamourt Goungel, Tamourt Boichiche, Tamourt Oum Lellé, Gâat Sawana and Tamourt Tali[1]) in the southern region of Hodh El Gharbi (Figure 11.1). Relative inaccessibility to the region (only one paved road) has stifled the collection of data at the national level, but the locations and variability of ephemeral wetlands are well known to local herders and farmers. Rainfall variability generally increases with encroaching aridity (Langbein, 1961; Williams, 1985) and this, in turn, controls the persistence, size and duration of ephemeral wetlands in the arid zone. Temporal variability is expressed using a coefficient of variation of inter-annual rainfall (CVIR). Areas with a CVIR of over 30% are considered to be non-equilibrium dynamic environments (Ellis *et al.*, 1994; Leach *et al.*, 1999). Average CVIR for weather stations in southern Mauritania (Ain Farba, Kobeni, Tamchekett, Tintane, Touil and Aïoun El Atrous) is at 43%, well above the threshold for non-equilibrium (Figure 11.2). As such, rainfall received in one wetland catchment can differ substantially from its neighbours and, indeed, the nearest weather station (Table 11.1). In addition,

---

[1] *Tamourts* are endorheic basins characterized by stands of *Acacia nilotica*, while *Gâats* form in shallower, more open depressions and are dominated by aquatic vegetation.

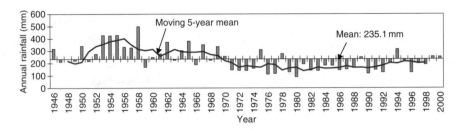

Figure 11.2 Mean annual rainfall at Aïoun (1946–2000)

Table 11.1 Annual rainfall data for southern Mauritania in 1999 and 2000

Weather stations	Annual rainfall (mm)		
	1999	2000	Difference
Aïoun El Atrous	255.4	253.1	−2.4
Ain Farba	289.3	479.0	189.7
Kobeni	519.6	363.2	−156.6
Tamchakett	138.2	230.1	91.8
Tintane	228.8	422.8	194.2
Touil	513.4	458.9	−54.4
Regional mean	324.1	367.8	43.7
Regional total	1944.7	2207.0	262.3

Source: Services Météorologique, Direction de l'Environnement et de l'Aménagement Rural (DEAR), 2000.

evapo-transpiration measurements range from 5.9 mm/day in Kiffa (west of the study region) to 8.5 mm/day in Néma (east of the study area), with peaks during the dry season (March–June). Such unpredictability tends to reduce confidence in rainfall instrument readings and opens the possibility for more spatially comprehensive and consistent measurements from remote sensing.

11.2.1   Data and processing

Data were collected at intervals spanning half a century, from aerial photographs taken in 1952–1956; from the Landsat-4 Thematic Mapper sensor in 1984–1985, and from GPS readings recorded in 1999–2000. Each set was used to assess annual and seasonal areal changes in five of the 244 ephemeral wetlands.

One of the earliest sources of reliable remote sensor data available for southern Mauritania is a set of panchromatic aerial photographs, commissioned and owned by the French IGN (Institut Géographique National). The aerial photographs were taken

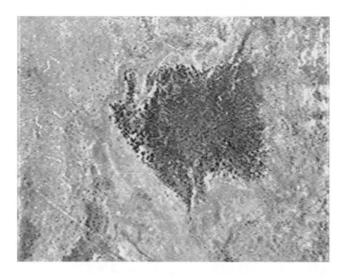

Figure 11.3 Aerial photograph of Oum Lellé. The ephemeral wetland is distinctive as an area of dark hue and coarse texture, representing biomass and water in the midst of an almost permanently arid and barren landscape. See Figure 11.3 for a low-altitude view

during 12–22 November 1956 (15 June 1952 for Tali) at a scale of approximately 1:50 000. The four November photographs were taken approximately 2–4 weeks after the end of the habitual wet season, when wetland surface areas are typically at their maximum levels; the photograph of Tali in 1952 was taken at the end of the dry season, when the surface area was much lower than the maximum. All wetlands are clearly visible on the aerial photographs as areas of much darker tone and coarser texture than their arid surroundings (Figure 11.3). Low solar reflectance off water, marked changes in vegetation healthiness and abundance between wet and dry soils, and changes in shading due to inundation and recent flooding, all contribute to sudden hue, tonal and textural changes that allow distinct boundaries to be routinely defined and measured. Tamourts are specific types of local wetlands, which are particularly discernible because of the characteristic presence of dense stands of *Acacia nilotica* trees (noticeable across Oum Lellé on the low-altitude photograph in Figure 11.4).

Two multispectral satellite images (path 201, row 049) were obtained from the Landsat-4 TM sensor; one coinciding with the wet season taken on 21 October 1984, the other during the dry season on 15 April 1985. Allowing for the usual caveats in change detection, both images are completely cloud-free, were taken at approximately the same time of day, and have similar near-equinox sun angles. The October 1984 wet season image is comparable with the aerial photographs taken during the wet season in November 1956, whilst the image captured during the dry season in April 1985 provides an opportunity to measure seasonal variations

Figure 11.4 Low-altitude photograph of Oum Lellé wetland in southern Mauritania

during a year that formed part of an unusually prolonged period of low rainfall in the region (Figure 11.2). This dry period had inevitable negative repercussions on the abundance and vitality of the natural vegetation, and further emphasized the abrupt changes in the environment between the wet/moist conditions and relatively abundant vegetation within wetlands and the dry/very sparse vegetation of the surroundings.

Translated into multispectral terms, distinct radiometric gradients between moisture and aridity should be observable using all seven channels of the Landsat-4 TM sensor data. One way to establish this distinctiveness between areas of vegetation associated with the wetlands and the surrounding aridity is by using the standard NDVI, which, in this chapter, is implemented using IDL®-modified programmes from the Research Systems Inc. ENVI® software (ENVI, 2003):

$$NDVI = (NIR_{band4} - VIS_{band3})/(NIR_{band4} + VIS_{band3})$$ (11.1)

The NDVI is a ratio of Landsat TM band 3 (red) and Landsat TM band 4 (near-infrared). In Figure 11.5, NDVI values representing a random cross-section (X-profile) of the Tamourt Tali wetland are clearly evident as a sharp peak of positive values, up and above the negative baseline of the surrounding aridity. This sharp peak (the black X-profile) is of NDVI values representing the wet season. NDVI values calculated from the Landsat TM image during the dry season for the same wetland are less conspicuous, resulting in a much lower, unobtrusive peak (the grey X-profile). Nevertheless, even these lower NDVI values are still capable of measuring the contrast between vegetation (regardless of abundance) associated with wetlands and the surrounding aridity of almost permanent lack of biomass. The two profiles in Figure 11.5 are based on a small sample of NDVI values for one

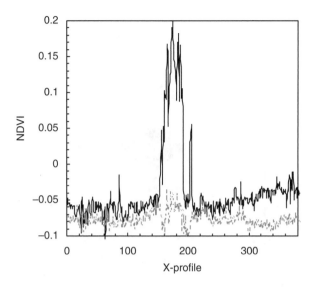

Figure 11.5 NDVI values of an X-profile across Tali during the wet season (black line) and dry season (grey line)

wetland. If all the NDVI values representing both wet and dry seasons are compared, it then becomes possible to visualize the total degree of variability. Figure 11.6 illustrates that the vast majority of the NDVI wet and dry season values are almost identical, underlining a highly stable, if somewhat sparse, distribution of vegetation. The two main differences are the greater proportion of pixels representing the wet season (NDVI values above 0, indicating a greater abundance of vegetation) and, conversely, the greater number of pixels representing the dry season (NDVI values below –0.2, characteristic of a scarcity in vegetation cover).

NDVI values are reliable surrogate measures for wetland areas in arid areas such as southern Mauritania, where the vast majority of dense vegetation cover is associated with some level of standing water. NDVI values are suitable for locating bodies of water in arid areas, but not for demarcating their exact spatial outlines. Instead, an iterative binary multispectral classification of the Landsat TM images into areas of water and non-water was necessary to augment the NDVI measurements. Statistically, the multispectral signature of standing water is highly distinctive, especially when compared to the high reflective properties of the arid surroundings. As a result, an unsupervised ISODATA clustering algorithm was sufficient to classify distinct pixels into a thematic class labelled as standing water. The supervised classification was implemented using ENVI® (RSI) software and modified through a series of iterative masking cycles of 12 classes at a time. Class separability statistics measured the distinctiveness of water bodies at the 90% level.

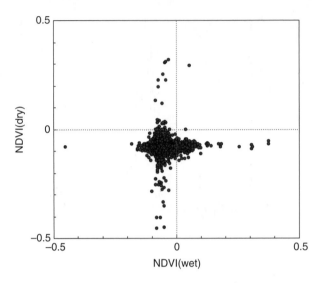

Figure 11.6 NDVI pixel values of Landsat TM images taken during the dry and wet seasons

Finally, a positional co-occurrence of high NDVI values and the water class, both derived from the Landsat TM images, resulted in the identification of all five ephemeral wetlands in both wet and dry seasons. It was then a straightforward matter of calculating the surface areas using descriptive statistics. The process was hindered only by areas of the images that represented vegetation associated with seasonal rivers. However, these areas were quickly identified (and subsequently eliminated) both by their narrow linear patterns (uncharacteristic of basin wetlands) and by their distance from the known location (based on local information) of the five ephemeral wetlands. Indeed, information on the spatial positioning of these established yet seasonal wetlands was further confirmed by the collection of ground-based coordinates, using a hand-held GPS receiver.

The most recent maps of the wetlands were created *in situ* by walking around the five in our study using a Garmin 2 plus GPS in 1999 and 2000 (Garmin, 1997). The GPS created track files based on records taken every 30 s. This method is well suited to mapping small, arid-zone wetlands but time consuming for large wetlands. The maximum extent (end of the wet season) of the wetlands was recorded in October–November 1999 and a minimal value (during the dry season) was recorded in March 2000. The track files created by the GPS were downloaded using Pcx5 and exported to ESRI®'s ArcGIS® (ArcGIS 2003), re-projected to the national coordinate system and then converted to polygons in order to calculate surface area.

The surface areas of the five ephemeral wetlands calculated using the four types of data are shown in Table 11.2. Maximum and minimum surface areas are shown for 1984–1985 and 1999–2000 using Landsat TM sensor data and GPS

Table 11.2 Surface areas (ha) of five ephemeral wetlands during maximum and minimum periods across five time intervals

Ephemeral wetland	Aerial photos	Landsat-4 thematic mapper		Differential GPS	
	Nov 1956 Max (ha)	Oct 1984 Max (ha)	Apr 1985 Min (ha)	Oct 1999 Max (ha)	Mar 2000 Min (ha)
Boichiche	23.45	1.04	0.00	203.64	8.65
Goungel	268.30	48.68	0.00	638.05	307.12
Oum Lellé	110.03	61.04	0.00	215.10	54.73
Sawana	176.38	47.64	0.00	834.77	16.42
Tali	*226.12	100.72	43.71	966.11	685.56
Average	144.54	51.82	8.74	571.53	214.50

*June 1952.

readings, respectively, but only maximum values are available from the 1952 aerial photographs.

11.2.2 Results

As expected, the smallest maximum and minimum surface areas of all five wetlands are measured by the Landsat TM images representing 1984 and 1985 respectively, years that correspond with a prolonged drought period (see Figure 11.1). Other than Tali, all have dried up completely by the dry season in April 1985, and even the maximum areas are far below those recorded by other data in 1956 and 1999 (Table 11.2). In contrast, the 1950s, represented by aerial photographs, is known as a period of above average rainfall (the 5 year average in 1956 was 402 mm/year). Although rainfall in Aïoun was lower (331.3 mm), it was still well above the 54 year average of 235.1 mm (Figure 11.2). This abundant rainfall is reflected in the larger average sizes (144.54 ha) of the five wetlands, especially in comparison to the 1985 average areas of 51.82 ha. However, the maximum surface areas of the same five wetlands in 1956 are generally smaller in size than in 1999 (average of 571.53 ha). This is contrary to annual precipitation averages recorded at Aïoun, which indicated that rainfall in 1999 (244.3 mm) was lower than in 1956 (331.3 mm). The largest maximum surface areas of any year were measured using GPS in 1999, surpassing those of 1956.

Variations in wetland surface areas also take place on a seasonal basis. Typically, wetlands in the Sahel reach their maximum surface areas by the end of the wet season, normally November. The volume of water then starts to decrease as water infiltrates into the substrate to the water table below and through evapo-transpiration, which increases rapidly during the dry season. Consumption by humans and live-stock will account for only minor outputs. Uncommonly, all five sample wetlands still held water, according to GPS measurements, in March 2000 (the dry season), and two of these, Goungel and Tali, even retained water until the following wet season. In contrast, seasonal variability is far more abrupt during the notorious drought years of the mid-1980s. In line with rainfall data, all of the wetlands were represented by Landsat TM sensor data as very small during the 1984 wet season; for example, Boichiche was only 1 ha in surface area. By the time of the wet season in 1985, all but Tali were measured as completely dry. Graphically, these boundary changes can be visualized in Figure 11.7. For completeness, the comparisons are only between the Landsat TM images in 1984–1985 and the GPS readings taken in 1999–2000; and for clarity the wetlands are not comparable in scale to each other. Generally, the spatial association of the boundaries is very close, giving support to the methods used to extract the information from each dataset. Shape and logical consistency (minimum boundaries are contained within maximum boundaries) are both preserved across space and time. Of the notable differences, there is a more dramatic decrease in surface area from wet season to dry season in Sawana and Biochiche, due to the shallow nature of the depressions. The shallowness in Sawana is further demonstrated by the fragmentation of the wetland after the relatively dry wet season of 1984. Maximum and minimum water level differences in Tali are less extreme due to the deeper nature of the wetland depression (on-site visits documented in Shine, 2002). It was not possible to assess seasonal variations in the 1950s due to the lack of comparable aerial photographs.

Comparisons between drought (1980s) and relatively non-drought (1956 and 1999) periods indicate a clear relationship between precipitation and wetland size. As a rule, below average rainfall means smaller surface areas, while above average rainfall results in larger surface areas. However, the disproportionately larger sizes of the wetlands with moderate rainfall of 255 mm in 1999–2000, compared to smaller surface areas in 1956 but with 331 mm of rainfall, needs further explanation. A number of hypotheses can be put forward to account for this disparity:

1. Errors arising from the use of different sources of data used to calculate surface areas.

2. Spatial variability in rainfall, resulting in less actual rainfall at the wetlands compared to recorded rainfall at the weather station at Aïoun.

3. Changes in the topography of the wetland basins over time.

Hypothesis (1) is an unavoidable factor in all multiscale and multitemporal analysis using remote sensing. Inconsistency in data sources is unavoidable, given the length of the study period and the isolated location of the study area. However, the effects of data source inconsistency are tolerable, provided that the scale and scope of measurement are within the objectives of the application. As alluded to earlier, the

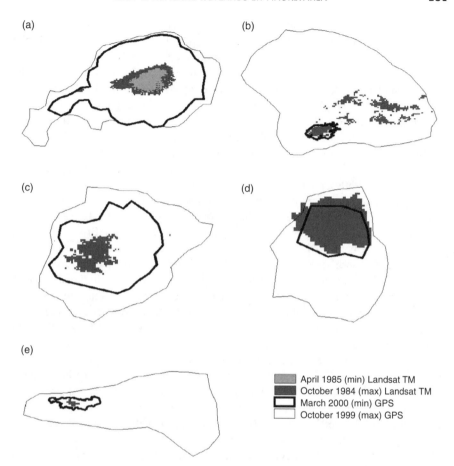

Figure 11.7 Min–max variations in wetland surface areas of (a) Tali, (b) Sawana, (c) Oum Lellé, (d) Goungel and (e) Boichiche, using Landsat TM (1984–1985) and GPS (1999–2000) data

scale of measurement is at a very fine level, the coarsest being from the Landsat TM sensor data at 30 m, and the scope is a straightforward identification of water bodies (and biomass associated with water) from highly arid surroundings. If errors are attributable to source data, they are more likely the result of manual interpretation of wetlands from aerial photographs, NDVI and classification imperfections from Landsat TM sensor data, and intrinsic positional errors from GPS measurements. However, given the magnitude of the wetland variations, both annually and seasonally, it seems highly unlikely that errors in source data are anything more than tangential.

Spatial variability of rainfall, hypothesis (2), may be more of a contributory impact on wetland size variations with distance from the nearest weather station.

The nearest weather station in Aïoun is at a distance of 10–140 km from the five wetlands. An extreme example can be demonstrated by Sawana and Oum Lellé; two wetlands a mere 7 km apart. Both wetlands filled to capacity in 1999 when the rainfall recorded in Aïoun was 253 mm. With similar precipitation recorded in the wet season of 2000, 255 mm. Sawana remained dry while Oum Lellé filled to levels comparable with 1999/00 (Shine, 2002). Such small-scale geographical differences may account for some changes but are less likely to account for variations in all five wetland surface areas across all years.

The third hypothesis seems the most likely explanation of the discrepancies between rainfall and wetland size. The drought that affected most of the Sahel in the 1970s, 1980s and nearly all of the 1990s resulted in reduced hydraulic erosion and the dominance of Aeolian erosion and deposition. Ephemeral wetlands in the arid zone are constantly changing in topographic shape, due to the interplay of such erosional and depositional forces (see e.g. Mullié and Brouwer, 1994). During the dry periods wetlands have been observed to infill, due to Aeolian deposition (Bouland, 1996). This is usually counterbalanced by the redistribution of sediment by hydraulic forces during the wet season. However, in the absence of sufficient rainfall during the wet season, Aeolian deposition continues to dominate and wetlands continue to infill. Communities in the Sahel frequently report sand encroachment as a major threat to wetlands (documented in PSB/GTZ, 1999). In time, sustained infilling tends to change the physical shape of the wetland basins, resulting in shallower depressions. If this is what happened to the sample wetlands during the 30 years or so of drought, the return of wetter conditions in 1999 resulted in a volume of water that filled areas greater than previous wetland depressions. This was corroborated through interviews with members of the communities living beside the sample wetlands in 2000, who claimed that the present wetlands are much larger than they had been in living memory.

The expansion of rain-fed agriculture and the accompanying clearing of land in the wetland catchments since the 1950s may also have contributed to the infilling of the wetland depressions. Encroaching agriculture on catchment slopes in the Sahel is known to increase run-off and sediment accumulation in the wetland depressions, thereby reducing wetland volume (Zimbabwe example in Whitlow, 1983). The consequence is shallower wetlands with reduced storage capacity (Mullié and Brouwer, 1994). This may have contributed to the enlargement of wetland surface areas in 1999–2000, when the wetlands overflowed their traditional depressions, causing flooding in nearby villages (e.g. Goungel in 1999).

Variability in wetland size and therefore in wetland duration is indisputable. That precipitation is a factor affecting wetland areas is also incontestable. However, other factors, such as erosion and deposition, land-use and human disturbance, may also have important roles to play. Conceptually, southern Mauritania is a non-equilibrium environment, based on dynamic ecologies (see Botkin, 1990) which, in turn, infer variability in space and in time. This means that environmental management systems based on average conditions and aiming to maintain the balance of nature are frequently unsuccessful, as they fail to understand the dynamics of non-equilibrium systems.

11.2.3   Implications for management

In simplest terms, the level to which a wetland fills in a year and the duration for which it holds water determines that year's productive capacity. As wetlands fill, they provide water for livestock and humid soil for flood-recession agriculture. In addition, wet conditions result in an abundance of wild foods and forestry products, which provide an increasingly important additional source of income to local communities. However, excess water can also have a negative effect on agriculture, as fields can remain flooded well into the growing season, increasing the risk of harvest failure – a delicate balance, therefore, between too much and too little water determines the annual productive capacity of the wetlands. Unfortunately, water levels in arid environments are impossible to predict, as they are dependant on highly variable rainfall. Instead, a tenuous situation prevails, where precipitation levels are frequently used at national level to gauge and evaluate annual wetland-derived resource availability.

Wetland management in the Sahel is traditionally based on the multiple use of resources by a wide range of consumers (see the Burkina Faso example in Bognounou *et al.*, 1994). Diversification assures security in the event of drought and minimizes the impact of human activities on the ecosystem. Unfortunately, natural resource managers and development project staff fail to understand the variable nature of these resources and implement strategies based on average or above-average conditions (e.g. PGRNP and MDRE, 1999). As there is no average or equilibrium state in dynamic environments, these goals are rarely met. Moreover, due to rainfall variability, a productive agricultural wetland may only be inundated for 1 in 3–4 years, with the implication that alternative sources of revenue are necessary in the intermediary years (e.g. Sawana in 2000–2001 and 2001–2002, when the wetland remained dry and crops absent). With the contemporary trend of converting wetlands from multiple-use to single-use systems comes the increase in economic risk during times of drought. Development agencies tend to base plans on Western principles, assuming equilibrium systems that are frequently unrealistic in arid areas (see Scoones, 1995). Mobile animal rearing and diversified livelihood strategies cope best in unpredictable situations, while arable agriculture is at the mercy of the rains. Until planners take the variability of resources in the arid zone into account, their interventions will be of minimal value, regardless of the vast quantities of capital injected into development schemes (observations in Shine, 2002).

The flora and fauna living in and using ephemeral wetlands have adapted to the variable conditions. Plants have resistant seeds and tubers (e.g. *Nymphea lotus*), and fish (such as *Clarias anguillaris* and *Protopterus annectens*) are equipped with lungs or resistant eggs to survive until the next rains. One particular newsworthy example is the rediscovery of relict populations of the Nile crocodile (*Crocodylus niloticus*) in 1999 (Nickel, 2001; Shine *et al.*, 2001), 70 years after the last reports of their existence and 6 years after the IUCN Species Survival Commission listed them as extirpated. These populations have survived dramatic environmental changes over the last 10 000 years by adapting to increasingly dry and variable conditions as the Sahara turned from savannah to desert. During the long dry seasons today, the crocodiles shelter in caves and burrows, waiting for the rains to return, bringing with them the hope of food in the form of fish, amphibians and small mammals. The isolated populations live on the brink, relying on the goodwill of the local populations and the sporadic rainfall for survival. Careful management practices, taking into account their reliance on these ephemeral ecosystems, are required to assure their future while we learn more about their distribution and survival tactics. Traditional land-use practices have preserved their habitats until now and formed a sound basis for future conservation efforts.

In addition, mobile species also make use of the wetlands; they provide staging posts and overwintering sites for a wide variety of Palaearctic migrants. Water birds have been counted in internationally important numbers in the ephemeral wetlands of south-eastern Mauritania, but this is not the case every year. When the wetlands remain dry, the birds go elsewhere, making it difficult to persuade conservationists and governments of their important role. Adaptations to the Ramsar Convention (Ramsar Convention Bureau, 1999) allow irregularly inundated wetlands in the arid zone to qualify as sites of international importance, based primarily on a minimum of 5 years of data. Unfortunately, data representing 5 years are generally unavailable in developing countries, and amendments for arid zone applications for Ramsar status have yet to be approved. Data from remote sensing, especially satellite sensors, are collected at frequent intervals, and certainly within the required annual basis. They have a potentially important role in routine monitoring of the presence and absence of wetlands, which, in turn, would contribute to migratory predictions. Wildlife and biodiversity managers need to take the unpredictable nature of these environments into account when drawing up management plans. Small, isolated habitat islands require priority conservation action, particularly when they host a wealth of biodiversity and/or rare species (see Pickett and Thompson, 1978; Pickett *et al.*, 1992).

11.3 Conclusions

Quality in strategies used for the management and development of ephemeral wetlands in the Sahel is critical for harmonizing ecological habitats with human necessity and economic sustainability. Without routine monitoring, the inherently variable nature of these arid zone wetlands makes this relationship even more crucial. The lack of information on the distribution and the severity of wetland variability in non-equilibrium systems inevitably leads to inappropriate and haphazard management schemes that are unavoidably associated with negative economic and environmental impacts. In response, this paper has explored the potential of adopting remote sensing for constructing methodologies that measure and compare the variability of wetland areas both seasonally and annually. Data from aerial photography, satellite imagery and GPS were used to delineate the minimum and maximum surface areas of a sample of ephemeral wetlands in southern Mauritania over a 44 year period. The delineation of arid zone ephemeral wetlands from remote sensor data is a matter of identifying distinctive combinations of water and vegetation from the surrounding permanent and almost barren landscape. The contrast is striking enough to produce highly accurate demarcations of wetland areas by photographic interpretation and unions of NDVI and water classification from satellite imagery. However, while the combination of remotely sensed data provide a comprehensive, multitemporal and relatively low-cost method of creating inventories of wetland resources in isolated areas, information on wetland characteristics and use is highly limited. Further information from field visits using GPS is necessary to not only provide a more detailed description of wetland profiles but also to aid the geo-registration and verification of the remotely sensed data.

Results suggest that the degree of annual and seasonal variability closely mirrors precipitation patterns. On the whole, larger wetland areas and narrower seasonal variability are measured by remote sensing and GPS during wetter years, whilst smaller areas and wider seasonal variability are represented during drier years. Some of the exceptions to this palpable rule are the result of data inconsistency and misinterpretation, as well as errors of rainfall interpolation between the sparse distribution of available weather stations. However, the most likely factor for deviations between wetland size and precipitation is basin shape. Increased Aeolian deposition due to the prolonged drought of the 1970s, 1980s and 1990s is the most likely cause of changes in basin shape in the Sahel. This, in time, has contributed to larger wetland surface areas in 1999 than in the 1950s, despite higher rainfall in the latter. It is anticipated that further work will focus on the use of digital elevation models, not only to measure basin shape and size but also to contribute to water depth calculations.

Ephemeral wetlands in the Sahel are by definition fleeting and highly variable. As such, associated potential agricultural productive capacity is so unpredictable that a harvest is rarely guaranteed. Historically, techniques centred on the multiple use of resources have been more resilient to wetland variability. Unfortunately, modern methods are more likely to be single-use, target-driven development schemes that assume optimal conditions. Such management plans, focused on one-off baseline studies or average conditions, are at best underproductive and at worst destructive when applied to dynamic ecosystems. Instead, development schemes and land-use or conservation planning should embrace the variability of ephemeral wetland resources and bear in mind that traditional multiple-use management systems have successfully sustained livelihoods and maintained biodiversity for centuries. Indigenous technical knowledge that draws on years of experience in managing non-equilibrium environments can form a sound basis for today's management of arid-zone ephemeral wetlands.

This chapter highlights the uniqueness and variability of ephemeral wetlands in the Sahel. It advocates a strong role for the implementation of remote sensing and GPS in long-term strategies that seek to survey their unpredictability as a precursor for balancing their ecological habitats with economic development. The monitoring of inland, arid-zone ephemeral wetlands in the developing world represents a unique application of remote sensing and a stark contrast to the majority of studies on far more permanent, temperate, coastal wetlands in the developed world. Unfortunately, the lack of research and inherent physical discontinuity has inevitably prevented the establishment of consistent ephemeral wetland taxonomies that are essential for long-term comparisons. Nevertheless, this chapter demonstrates how a mixture of data from various remote sensing sources can at least be used to measure broad wetland/non-wetland dichotomies whenever such data are available. Sporadic surveillance may not be ideal, but it does provide a basis for the establishment of benchmarks. Space-borne data, in particular, represent a source of routine regional scale surveillance at reasonable cost, which when compared across time can aid the consistent and accurate monitoring of the variability of arid zone ephemeral wetlands.

Acknowledgements

The authors would like to thank Dr Suzanne McLaughlin from the School of Environmental Sciences, University of Ulster, for her GIS help, and Project GIRNEM of the German Technical Cooperation (GTZ) for support in the field.

References

Anderson, J. R., Hardy, E. E., Roach, J. T. and Witmer, R. E. (1976) *A Land Use and Land Cover Classification System for Use with Remote Sensor Data.* US Geological Survey Professional Paper No. 964. US Government Printing Office: Washington, DC; 28 pp.

Bouland, P. (1996) *Rapport de mission dans la commune de Male, Wilaya de Brakna, pour l'établissement d'un projet d'appui a la coopération décentralisée avec la commune de Male.* Commission de l'Union Européenne: Nouakchott, Mauritania.

Ceccato, P., Flasse, S. and Grégoire, J.-M. (2002) Designing a spectral index to estimate vegetation water content from remote sensing data. Part 2: validation and applications. *Remote Sensing of Environment* **82**, 198–207.

Chopra, R., Verma, V. K. and Sharma, P. K. (2001) Mapping, monitoring and conservation of Harike wetland ecosystem, Punjab, India, through remote sensing. *International Journal of Remote Sensing* **22**, 89–98.

Cowardin, L. M. and Myers, V. I. (1974) Remote sensing for the identification and classification of wetland vegetation. *Journal of Wildlife Management* **38**, 308–314.

Dottavio, L. C. and Dottavio, D. F. (1984) Potential benefits of new satellite sensors to wetland mapping. *Photogrammetric Engineering and Remote Sensing* **50**, 599–606.

Dwiveldi, R. S. and Sreenivas, K. (2002) The vegetation and waterlogging dynamics as derived from spaceborne multispectral and multitemporal data. *International Journal of Remote Sensing* **23**, 2729–2740.

Ellis, J. E., Coughenour, M. B. and Swift, D. M. (1994) Climate variability, ecosystem stability, and the implications for range and livestock management. In Behnke, R. H. and Scoones, I. (eds), *Range Ecology at Disequilibrium:New Models of Natural Variability and Pastoral Adaptation in African Savannas*. Overseas Develeopment Institute: London, UK; 31–41.

Farinha, J. C., Costa, L. T., Zalidis, G., Mantzavelas, A., Fitoka, E., Hecker, N., and Tomas Vives, P. (1996) Mediterranean wetland inventory: habitat description system. MedWet/Instituto da Conservaçao da Natureza (ICN)/Wetlands International/Greek Biotope/Wetland Centre (EKBY): Lisbon, Portugal.

Guenda, W., Kabre, R., Ouédraogo, R. L. and Zongo, F. (1994) Ecologie et biodiversité des zones humides. In Sally, L, Kouda, M. and Beaumond, N. (eds), *Zones Humides du Burkina Faso*. IUCN: Gland, Switzerland, 81–112.

Harvey, K. R. and Hill, G. J. E. (2001) Vegetation mapping of a tropical freshwater swamp in the Northern Territory, Australia: a comparison of aerial photography, Landsat TM and SPOT satellite imagery. *International Journal of Remote Sensing* **22**, 2911–2925.

Howland, W. G. (1980) Multispectral aerial photography for wetland vegetation mapping. *Photogrammetric Engineering and Remote Sensing* **46**, 87–99.

Jensen, J. R., Hodgson, M. E., Christensen, E., Mackay, H. E. Jr, Tinney, L. R. and Sharitz, R. R. (1986) Remote sensing inland wetlands: a multispectral approach. *Photogrammetric Engineering and Remote Sensing* **52**, 87–100.

Jensen, J. R., Ramsay, E. W., Mackay, H. E. Jr, Christensen, E. J. and Sharitz, R. R. (1987) Inland wetland change detection using aircraft MSS data. *Photogrammetric Engineering and Remote Sensing* **53**, 521–529.

Kingsford, R. T. (1995) Occurrence of high concentrations of waterbirds in arid Australia. *Journal of Arid Environments* **29**, 421–425.

Kingsford, R. T. (1997) *Wetlands of the World's Arid Zones*. Ramsar Convention Bureau: Gland, Switzerland.

Leach, M., Mearns, R., and Scoones, I. (1999) Environmental entitlements: dynamics and institutions in community-based natural resource management. *World Development* **27**, 225–247.

Lee, K. H. and Lunetta, R. S. (1995) Wetlands detection methods. In Lyon, J. G. and McCarthy, J. (eds), *Wetlands and Environmental Applications of GIS*. CRC Press: Boca Raton, FL, USA, 245–250.

Lyon, J. G. (2001) *Wetland Characterization: GIS, Remote Sensing and Spatial Analysis*. Taylor and Francis: London, UK.

Manson, F. J., Loneragan, N. R., McLeod, I. M. and Kenyon, R. A. (2001) Assessing techniques for estimating the extent of mangroves: topographic maps, aerial photographs and Landsat TM images. *Marine Freshwater Research* **52**, 787–792.

Maselli, F., and Rembold, F. (2002) Integration of LAC and GAC NDVI data to improve vegetation in semi-arid environments. *International Journal of Remote Sensing* **23**, 2475–2488.

Mullié, W. C., Brouwer, J., and Scholte, P. (1994) Numbers, distribution and habitat of wintering white storks in the east-central Sahel in relation to rainfall, food and anthropogenic influences. International Symposium on the White Stork (Western Population), Basel, Switzerland.

Mumby, P. J., Green, E. P., Edwards, A. J., and Clark, C. D. (1999) The cost-effectiveness of remote sensing for tropical coastal resources assessment and management. *Journal of Environmental Management* **55**, 157–166.

Nohara, S. (1991) A study on annual changes in surface cover of floating-leaved plants in a lake using aerial photography. *Vegetation* **97**, 125–136.

PGRNP (Projet de Gestion des Resources Naturelles en Zone Pluviale) and MDRE (Ministère de Développment Rurale). (1999) Diagnostique du terroir villageois de Goungel. PGRNP: Nouakchott, Mauritania.

Pickett, S. T. A. and Thompson, J. N. (1978) Patch dynamics and the design of nature reserves. *Biological Conservation* **13**, 27–37.

Pratt, N. D., Bird, A. C., Taylor, J. C. and Carter, R. C. (1997) Estimating areas of land under small-scale irrigation using satellite imagery and ground data for a study area in N. E. Nigeria. *The Geographical Journal* **163**, 65–77.

Roux, F. and Jarry, G. (1984) Numbers, composition and distribution of populations of Anatidae wintering in West Africa. *Wildfowl* **35**, 48–60.

Rutchey, K and Vilcheck, L. (1994) Development of an everglades vegetation map using a SPOT image and the global positioning system. *Photogrammetric Engineering and Remote Sensing* **60**, 767–775.

Scoones, I. (ed.). (1995) *Living with Uncertainty. New Directions in Pastoral Development in Africa.* Intermediate Technology Publications: London, UK.

Shaikh, D., Green, D. and Cross, H. (2001) A remote sensing approach to determine environmental flows for wetlands of the Lower Darling River, New South Wales. *Australia International Journal of Remote Sensing* **22**: 1737–1751.

Shine, T., Böhme, W., Nickel, H., Thies, D. F. and Wilms, T. (2001) Rediscovery of relict populations of the Nile crocodile *Crocodylus niloticus* in south-eastern Mauritania, with observations on their natural history. *Oryx* **35**, 260–262.

Shine, T. (2002) An Integrated Investigation of Ephemeral Wetlands in Eastern Mauritania and Recommendations for Management. PhD thesis, School of Biological and Environmental Sciences, University of Ulster at Coleraine, Northern Ireland.

Shultz, G. A. and Engman, E. T. (eds). (2000) *Remote Sensing in Hydrology and Water Management.* Springer: Berlin.

Simmons, R. E., Barnard, P. E. and Jamieson, I. G. (1999) What precipitates influxes of birds to ephemeral pans in arid landscapes? Observations from Namibia. *Ostrich* **70**, 145–148.

Stevenson, N. and Frazier, S. (1999) Review of wetland inventory information in Africa. In Finlayson, C. M. and Spiers, A. G. (eds), *Global Review of Wetland Resources and Priorities for Wetland Inventory.* Wetlands International: Wageningen, The Netherlands.

Stewart, W. R., Carter, V. and Brooks, P. D. (1980) Inland (non-tidal) wetland mapping. *Photogrammetric Engineering and Remote Sensing* **46**, 617–628.

Taylor, A. R. D., Howard, G. W. and Begg, G. W. (1995) Developing wetland inventories in Southern Africa: a review. *Vegetation* **118**, 57–79.

Turner, R. K., Georgiou, S., Brouwer, R., Bateman, I. J. and Langford, I. J. (2003) Towards an integrated environmental assessment for wetland and catchment management. *The Geographical Journal* **169**, 99–116.

Williams, D. C. and Lyon, J. G. (1997) Historical aerial photographs and a geographic information system (GIS) to determine effects of long-term water level fluctuations on wetlands along the St. Marys River, Michigan, USA. *Aquatic Botany* **58**, 363–378.

Williams, W. D. (ed.) (1998) *Wetlands in a Dry Land: Understanding for Management.* Environment Australia, Biodiversity Group: Canberra, Australia.

Index